04-CTS-3T

Evaluation of Bacterial Pathogen and Indicator densities After Dewatering of Anaerobicall Digested Biosolids Phase II and III

by:

Matthew J. Higgins, Ph.D.
Bucknell University

Yen-Chih Chen, Ph.D.
Bucknell University

Don Hendrickson, Ph.D.
Hoosier Microbiology Laboratory

Sudhir N. Murthy, Ph.D., P.E.
DCWASA

2008

WERF

The Water Environment Research Foundation, a not-for-profit organization, funds and manages water quality research for its subscribers through a diverse public-private partnership between municipal utilities, corporations, academia, industry, and the federal government. WERF subscribers include municipal and regional water and wastewater utilities, industrial corporations, environmental engineering firms, and others that share a commitment to cost-effective water quality solutions. WERF is dedicated to advancing science and technology addressing water quality issues as they impact water resources, the atmosphere, the lands, and quality of life.

For more information, contact:
Water Environment Research Foundation
635 Slaters Lane, Suite 300
Alexandria, VA 22314-1177
Tel: (703) 684-2470
Fax: (703) 299-0742
www.werf.org
werf@werf.org

This report was co-published by the following organization. For non-subscriber sales information, contact:

IWA Publishing
Alliance House, 12 Caxton Street
London SW1H 0QS, United Kingdom
Tel: +44 (0) 20 7654 5500
Fax: +44 (0) 20 7654 5555
www.iwapublishing.com
publications@iwap.co.uk

Library of Congress Catalog Card Number: 2007921817
IWAP ISBN: 978-1-84339-784-7

Bucknell University, District of Columbia Water and Sewer Authority

ACKNOWLEDGMENTS

The project team thanks WERF for its support of the research, and the District of Columbia Water and Sewer Authority (DCWASA) for additional funding and support of the project. In addition, the team thanks the many personnel at facilities that helped with sampling and provided numerous hours of in-kind contributions to this project. The project team thanks the project subcommittee members for their valuable guidance and oversight. Finally, thanks go to WERF program managers for their support and assistance.

Report Preparation

Principal Investigators:

Matthew J. Higgins, Ph.D.
Yen-Chih Chen, Ph.D.
Bucknell University

Don Hendrickson, Ph.D.
Hoosier Microbiology Laboratory

Sudhir N. Murthy, Ph.D., PE
DC Water and Sewer Authority

Project Advisors:

Joseph Farrell, Ph.D., PE

Environmental Consultant

Perry Schafer, PE, DEE

Brown and Caldwell

Project Collaborators:

Christopher Peot

DC Water and Sewer Authority

William Toffey

Philadelphia Water Department

Project Subcommittee:

Water Environment Research Foundation Staff

ABSTRACT AND BENEFITS

Abstract:

The increase in culturable densities (ICD) of indicator bacteria, mainly fecal coliforms (FCs) and *E. coli*, immediately following dewatering, was evaluated at several full-scale facilities, in addition to the increases measured during cake storage. The results showed that the increase immediately after dewatering was a statistically verifiable occurrence at some facilities, but not all, as was the additional increases measured during cake storage. The immediate increase and growth were much more prevalent in processes that utilized centrifuge dewatering compared to belt filter press dewatering. In addition, thermophilic digestion processes typically experienced much greater increases immediately after dewatering compared to mesophilic processes. For example, increases of up to five orders of magnitude were measured in thermophilic processes. In comparison, the immediate increase after dewatering was typically 0 – 1 order of magnitude for mesophilic processes. Only one plant was sampled with high solids centrifugation that did not show increases immediately after dewatering and after storage and this plant was different from others in that it utilized thermophilic anaerobic digesters in series. The results showed a good correlation between the digestion temperature and the magnitude of increase measured after dewatering. As temperature of digestion increased, the magnitude of the increase also increased. The digestion SRT and VS reduction did not correlate well with the magnitude of increase after dewatering and storage. Sampling of several thermophilic Class A plants and a thermophilic Class B plant showed that bacterial pathogens, such as *Salmonella,* do not increase after dewatering and storage. In comparison, increases in *Salmonella* densities were observed during cake storage at two Class B mesophilic plants.

Benefits:

- Develops a database of utilities with and without fecal coliform increases which can be used for evaluation of this phenomenon.
- Develops a standardized protocol for evaluating increases in inidicator bacteria after dewatering.
- Develops an approach to measure the potential for increases using a combination of quantitative polymerase chain reaction (qPCR) and standard culturing methods for enumerating indicator bacteria, namely FC and *E. coli.*
- Demonstrates that *Salmonella* do not increase after dewatering of Class A biosolids and a Class B thermophilically digested biosolids.
- Describes the potential hypothesis to describe the mechanisms for the increases.
- Evaluates the process parameters that contribute to increases in FC densities.

TABLE OF CONTENTS

LIST OF TABLES

LIST OF FIGURES

LIST OF ACRONYMS

AI	autoinducer
ATAD	autothermal thermophilic aerobic digestion
BSA	bovine serum albumin
CTAB	hexadecylmethylammonium bromide
cPCR	competitive polymerase chain reaction
DNA	deoxyribonucleic acid
DS	dry solid
EDTA	Ethylenediamine-N,N,N',N'-tetraacetic acid
FC	fecal coliform
gadA/G	glutamate decarboxylase gene
HML	Hoosier's Microbiological Laboratory
HPC	heterotrophic plate count
LSD	least significant difference
MF	membrane filtration
MGD	million gallons per day
MPN	most probable number
PCR	polymerase chain reaction
qPCR	quantitative polymerase chain reaction (such as real time PCR)
RT-PCR	real time PCR
RNA	ribonucleic acid
RNC	reversibly non-culturable
SCM	standard culturing method
SDS	sodium dodecyl sulfate
SI	sudden increase
SM	standard methods
SRT	solids retention time
TBE	tris-borate-EDTA
TD	thermophilic digester
TE	tris-EDTA
TMC	total microbial counts
TPAD	temperature phased anaerobic digestion
VBNC	viable but non-culturable
VSR	volatile solids reduction

EXECUTIVE SUMMARY

Recent literature reports as well as the previous WERF funded project reported that a sudden increase in indicator bacteria, mainly fecal coliform and/or *E. coli*, was being measured immediately after dewatering compared to just prior to dewatering. Depending on the process, the increase could be 1-5 orders of magnitude. In addition, storage of cakes led to additional increases in the indicator density of up to several orders of magnitude within a few days of storage. During storage, a peak density was observed within a few days, followed by a decrease in densities with additional storage. The increase immediately after dewatering has been termed "sudden increase" or SI while the increase during cake storage has been termed "regrowth".

Several important research needs were identified related to the sudden increases after dewatering and increases during cake storage, and several of the research needs evolved into the main objectives of this project. The main objectives of this research were focused on understanding the sudden increase after dewatering as well as the increase measured during storage, and especially to:

- examine the impact of different process parameters such as reactor configuration, digestion temperature, solids reduction, solids retention times, dewatering equipment on the extent of the increases;
- determine if potentially pathogenic bacteria, such as *Salmonella*, also increased after dewatering of Class A and Class B biosolids.
- examine potential mechanisms to explain the sudden increase after dewatering followed by regrowth;

The project team performed sampling at nine full-scale facilities with the combination of anaerobic digestion and either centrifuge (both high and low solids) or belt filter press dewatering. The digestion processes included multi-stage and single stage thermophilic, pre-pasteurization followed by mesophilic, and conventional mesophilic. Samples were collected from the different stages of the process, and generally analyzed for fecal coliforms and *E. coli* using standard culturing methods (SCMs), and for *E. coli* density using molecular methods which enumerate the number of *E. coli* DNA copies present in the sample. This method, quantitative polymerase chain reaction (qPCR), was used to determine if the indicator bacteria were potentially being underestimated by SCM enumeration. For the plants that used centrifuge dewatering, all experienced SI and/or regrowth except one plant which had thermophilic reactors in series. The three plants with belt filter press dewatering (and conventional mesophilic digestion) did not have significant sudden increases and/or regrowth. The results also indicated that as the temperature of digestion increased, the SI potential also increased. In other words, for thermophilic digestion, the difference between qPCR and SCM increased, but the densities measured by qPCR remained in the range of 10^4-10^5 per gram dry solids. Therefore, the immediate increase potential measured as the difference between the qPCR and the SCM density, was greater for thermophilic processes. In addition, the change in the SCM measured density after dewatering was greater for the thermophilic processes compared to the mesophilic processes. Mesophilic processes often had a 0-1 order of magnitude increase, while thermophilic processes could have an immediate increase of 2 to 5 orders of magnitude after centrifuge dewatering. During cake storage, the thermophilic processes typically had an additional increase between about 2 and 3.5 $\log_{10}$ in the FC density, whereas the mesophilic processes had an increase in FC density between 0.5 and 2.5 $\log_{10}$. Operational parameters such as VS reduction or digestion SRT were not well correlated with the SI potential or regrowth. The results also showed a very good positive linear correlation between the increase potential and the actual increase that occurred. This finding helps support the hypothesis that the SCM underestimated the indicator bacteria density prior to dewatering, and dewatering resulted in a increased enumeration by the culturing method.

An evaluation of data from the literature and utilities was performed to broaden the scope of data available for analysis. Although some of these reports did not examine regrowth, nor provide enough data for statistical analysis, the results do provide some useful information. Plants that utilized the combination of digestion and centrifuge dewatering had a high probability of experience either SI and/or regrowth – about 30 out of 35 plants or 86%. In comparison, for plants utilizing digestion with belt filter press dewatering, one out of nine reported increases immediately after dewatering and/or regrowth.

Laboratory data were collected to examine the effect of time and temperature on bacterial decay and the increase potential. As the SRT increased for anaerobic mesophilic digestion, the FC densities measured by culturing methods also decreased and the result agreed well with full-scale data. The qPCR results showed that as SRT increased, the difference between the qPCR and SCM densities also increased, and was about a one order of magnitude difference at SRTs greater than about 20 days. This increased potential measured in the laboratory reactors was similar to the field results with mesophilic anaerobic digestion. Laboratory thermophilic digestion also showed similar results to the field, in that as time and temperature increased, the difference between the qPCR and SCM densities also increased. These differences, or increase potential was up to 5 orders of magnitude, and this was similar to the field results.

Preliminary studies were performed to investigate the development of a laboratory method that would simulate reactivation and regrowth. This method was seen as critical for additional research on mitigation strategies such as digestion time-temperature experiments. Several studies that examined media additives and cell washing were not very successful at reactivating the non-culturable organisms. Two experiments that utilized a lab process developed at Bucknell University to simulate a high solids centrifuge process was able to produce a cake that had a similar regrowth curve as samples from a full-scale centrifuge.

Three Class A plants with anaerobic digestion, one Class B with thermophilic anaerobic digestion, and two Class B plants with mesophilic anaerobic digestion were sampled for a number of different potentially pathogenic bacteria. One of the Class A plants was sampled multiple times. Using *Salmonella* as the key bacterial pathogen to consider, none of the Class A plants had reactivation or regrowth of *Salmonella* after centrifuge dewatering, including the multiple samples taken from one of the Class A plants. The thermophilic anaerobic digestion plant considered Class B also did not have reactivation and/or regrowth of *Salmonella*. The two mesophilic anaerobic digestion processes with centrifuge dewatering had regrowth of *Salmonella* in the cake during storage.

Several different theories have been postulated by researchers to explain the sudden increase in culturable densities after dewatering, which include:

1. Release of bacteria due to floc breakup during dewatering which allows more accurate enumeration;
2. False positives associated with enumeration techniques;
3. Regrowth of bacteria after dewatering;
4. Removal of growth inhibitors during dewatering;
5. Contamination from the centrifuge;
6. Reactivation of non-culturable bacteria.

Research has been performed to examine most of these theories, either by this research team or others. Based on the literature and data from this study, it appears that, in some cases, the standard culturing method can underestimate the densities after digestion and prior to dewatering. Immediately following dewatering, the densities of culturable indicator bacteria is a result of better or more accurate enumeration by the standard culturing method. Based on this research and work of others, the Principle Investigators of this project believe that the sudden increase after dewatering is largely due to reactivation of bacteria that had become non-culturable during digestion. The increased during cake storage are believed to be a result of regrowth.

In summary, the results from this work support the conclusion that the sudden increase in FC and *E. coli* after dewatering followed by regrowth during cake storage is an actual, statistically verifiable occurrence at some plants. This sudden increase followed by additional growth has been previously termed "reactivation and regrowth". The combination of digestion with high solids centrifuges appears to increase the likelihood of this occurrence. The combination of digestion with belt filter press has a lower probability of reactivation and regrowth. Pathogenic bacteria, such as *Salmonella,* did not "suddenly increase" or regrow in the Class A biosolids sampled.

CHAPTER 1.0

INTRODUCTION

1.1 Overview of Sudden Increase of Fecal Coliform Density after Dewatering

Over the last few years, utilities have reported substantial and sudden increases in the density of indicator bacteria after dewatering compared to just prior to dewatering. In addition, storage of the biosolids after dewatering led to additional increases. The indicator bacteria included fecal coliforms (FCs) and *E. coli*. The processes reporting this sudden increase were typically employing anaerobic digestion (both mesophilic and thermophilic) and centrifuge dewatering. For example, Figure 1-1 shows the mean fecal coliform densities of triplicate samples measured before and after thermophilic digestion, after centrifuge dewatering, and after on-site storage in a silo. Prior to digestion, the FC density is approximately 10^7 per gram dry solids (DS). After single stage, thermophilic digestion ($\approx$ 15 days at 55°C), the density was reduced to about 10^2 per gram DS as measured just prior (after storage) to the centrifuge. Immediately after centrifuge dewatering, the density increased several orders of magnitude to almost 10^5 per gram DS, and the density increased further when the cake was stored on site.

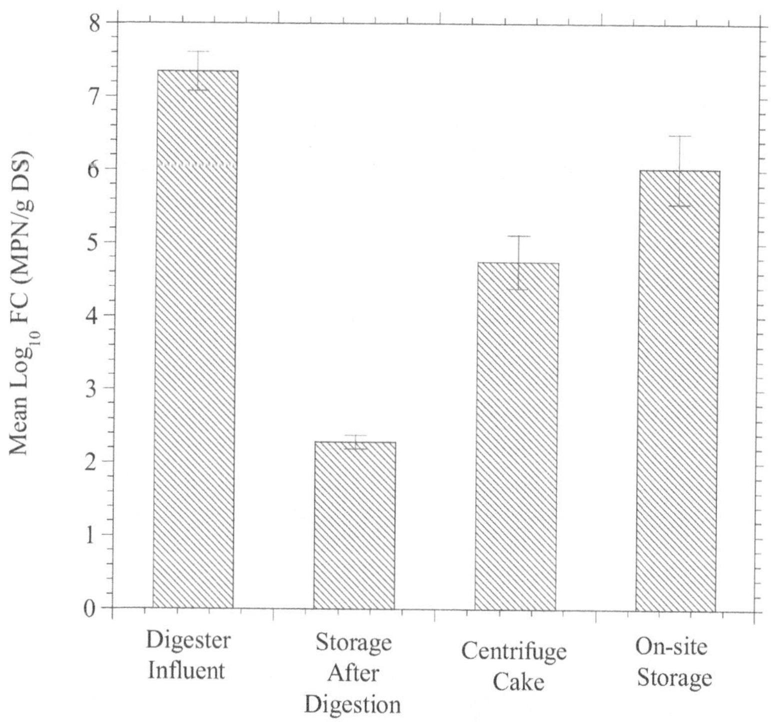

Figure 1-1. FC Counts from Different Locations Before and After Thermophilic Digestion, After Dewatering, and After On-Site Cake Storage.

This sudden increase immediately after dewatering followed by additional increases during storage shown in Figure 1-1 has also been reported in the literature by several utilities and researchers (Cheung et al., 2003, Qi et al., 2004, and Monteleone et al., 2004, Erdal et al., 2003 and 2004, Iranpour et al., 2002 and 2003, Hendrickson et al., 2004). These papers are reviewed here to provide additional historical information on this phenomenon.

In August of 2001 the largest known installation of a temperature phased anaerobic digestion process at a 60-MGD wastewater treatment plant was placed into operation. The facility met the time and temperature requirement for Class A biosolids. Samples collected after the digestion process, which included thermophilic/anaerobic followed by mesophilic/anaerobic digestion, indicated there was no detectable level of FC bacteria in the treated biosolids. However, subsequent testing of the biosolids following dewatering by high-solid centrifugation revealed high levels of fecal coliform bacteria. These dewatered biosolids did not meet the Class B requirement for biosolids.

On August 23, 2001, with the help of personnel at the facility, numerous samples were collected aseptically, stored on ice and transported by air to Dr. Hendrickson's laboratory, Hoosier Microbiological Laboratory, Inc. (HML), for testing (Hendrickson, 2001). The results of the study are shown in Table 3-10.

This study indicated that a large increase in FC density was occurring following high-solids centrifugation. The dramatic appearance of large numbers of FC could not have been the result of cross-contamination (samples were collected aseptically immediately downstream of dewatering device) or false positives since only 10% of the isolates were identified as non-coliforms or false positives. The bacterial isolates were identified by Biolog^{TM}. Biolog^{TM} tests the ability of a microorganism to utilize or oxidize a preselected panel of different carbon sources. The test yields a characteristic pattern which constitutes a metabolic fingerprint of the capacities of the unknown organism thereby allowing us to identify the bacterium.

In a separate test, the high-solids centrifuges were disinfected using a sodium hypochlorite solution prior to operating it with the digested solids. Immediately after disinfection, samples of digested cake were taken downstream of the centrifuge to be analyzed for regrowth. Fecal regrowth was observed in all the samples. This test shows that seeding of the cake from contamination from centrifuges was likely not occurring.

All bacterial samples were stored on ice and processed within 24 hours at HML. Therefore, all samples met the 24-hour limit established by the U.S. EPA. In addition duplicate samples were processed at the wastewater treatment lab within six hours after collection for total coliform analysis by membrane filtration. Dr. Hendrickson observed the sample analysis by the wastewater lab, which had similar regrowth results to those of HML.

A very small continuous feed of 15 ppm (v/v) hypochlorite in digested solids downstream of the high solid centrifuge destroyed a majority of microorganisms and produced a Class B biosolids. This is of considerable interest, since this small dose was not expected to cause much harm to these microorganisms.

Table 1-1. Fecal Coliform Density of Biosolids Collected at a Plant with Thermophilic/Anaerobic Digestion and High-Solid Centrifugation.

Location and Time of Sampling	Fecal Coliform Density in Dry Solids
Primary untreated sludge	
Collected at 11:45 AM	85,106,382/g dry
Collected at 3:45 PM	47,619,047/g dry
Digester sludge discharge (anaerobic/thermophilic)	
Collected at 1:12 PM	<6/g dry
Collected at 2:15 PM	<8/g dry
Digester sludge discharge (anaerobic/mesophilic)	
Collected at 1:05 PM	<12/g dry
Collected at 2:30 PM	<12/g dry
Digester sludge secondary (floating cover/aerobic/mesophilic)	
Collected at 12:55 PM	<5/g dry
Collected at 2:40 PM	<6/g dry
Dewatering polymer	
Collected at 12:37 PM	<21/g dry
Collected at 3:00 PM	<21/g dry
Feed to high solid centrifuge	
Collected at 12:45 PM	<4/g dry
Collected at 2:55 PM	<3/g dry
Centrifuge discharge	
Collected at 12:30 PM	381,174,640/g dry
Collected at 3:00 PM	19,893,428/g dry
Cake pump discharge to storage hopper	
Collected at 12:10 PM	220,690,000/g dry
Collected at 3:30 PM	36,500,000/g dry
Truck sludge	
Collected at 11:25 AM	61,456,035/g dry
Collected at 2:45 PM	85,419,734/g dry
Sludge placed on drying beds for 19 days	
Surface sample	2,454,395/g dry
Core (6-8" deep)	140,943/g dry
Crust	374,532/g dry

Hendrickson (2001) was one of the first reports of SI/regrowth, and since that time, several publications have also reported similar results. These reports are summarized in the following paragraphs. It should be noted that for the purpose of this review, "SI" is defined as an increase in FC or *E. coli* density in the cake solids collected immediately after centrifugation or other dewatering process compared to the feed to the dewatering equipment. "Regrowth" is defined as an additional increase in FC or *E. coli* density during further storage of the cake over a period of hours or days. Not all studies examined both issues.

Erdal et al. (2003) reported that SI and regrowth occurred from a mesophilic digested solids after both low solids and high solids centrifugation, however, after belt filter press dewatering, no SI occurred, and moderate regrowth after one day of storage. Interestingly, they also found that screw conveyance of the cake increased the SI and regrowth of the cake. Erdal et al. (2003 and 2004) found that a low dose lime application to the cake controlled SI and regrowth, especially when the pH was increased to greater than 8.5. They considered low dosages in the range of 1-9% lime on a dry solids basis, and the lime was added prior to the screw conveyance system.

Iranpour et al., (2002 and 2003) reported that SI and regrowth occurred after high solids centrifugation of a thermophilically digested biosolids. Prior to dewatering, the biosolids met Class A requirements of less than 1000 MPN/g DS of FC, but after dewatering, significant increases were measured in the FC density. Interestingly, Iranpour et al. (2002) found that lab centrifugation of the liquid solids did not result in SI or regrowth, suggesting that the g-force alone is not the main factor in causing SI and regrowth. They also reported that insulating the post digestion processes to maintain temperatures during dewatering and handling eliminated SI and regrowth (Iranpour et al., 2003).

Monteleone et al. (2004) examined five plants in the U.K. with mesophilic digestion. Four of the plants utilized centrifuge dewatering and one used belt filter press dewatering. All four sites using centrifuge dewatering measured an increase in *E. coli* density in the cake compared to the centrifuge feed, and two were statistically significant. The increases were reported as 63%, 74%, 394%, and 3452%. The one plant with belt filter press dewatering had a 44% decrease in *E. coli* density in the cake compared to the feed. No regrowth measurements were made as part of this study, and unfortunately, the data was not presented in such a way that the log increase could be determined. Similar to Iranpour (2002) and this WERF project, centrifuging the samples with a lab centrifuge at various g-values did not cause SI.

Cheung et al. (2003) examined a plant in the U.K. with mesophilic digestion and high solids centrifuge dewatering and found that after dewatering the *E. coli* density increased by up to 2.17 $\log_{10}$ units compared to the centrifuge feed. They also reported that several laboratory homogenization techniques did not improve the recovery of *E. coli* in the centrifuge feed, although an increase in the stomacher time from 2 to 4 minutes increased the recovery by about 0.5 $\log_{10}$ in cake samples.

Qi et al. (2004) reported results from four different plants in the U.S., all with mesophilic digestion. For one plant with a belt filter press, low solids centrifuge, and high solids centrifuge, SI did not occur after belt filter press dewatering, but an increase was measured after low and high solids centrifugation. At the second plant, they sampled eight days during a one month period, and for seven of the days, an increase in FC density was measured in the cake immediately after dewatering compared to the feed to the centrifuge, and for three of those days the increase was statistically significant. Eight samples were also collected at a third plant, and none of the samples had greater FC density in the centrifuge cake compared to the feed. At the fourth plant, they reported that statistically significant increases in FC density occurred for both low and high solids centrifuges from two different mesophilically digested biosolids. In addition to this SI, significant regrowth also occurred. Qi et al. (2004) also reported that the increase was not likely due to false positives, in that the presence of *Bacillus* sp. did not interfere with the enumeration results. Similar to Cheung et al. (2003), shear of the liquid digested product in the laboratory, in this case using a kitchen blender, did not result in SI.

In a WERF-funded study, Higgins et al. (2006) sampled seven different plants, several multiple times. The plants included mesophilic digestion, thermophilic digestion, and temperature phased anaerobic digestion (TPAD) which utilizes thermophilic followed by mesophilic digestion. Five of the seven plants had SI and regrowth. One thermophilic plant that was the only one to use thermophilic reactors in series appeared did not have SI and regrowth. One TPAD plant that did not have SI and regrowth had a very high industrial flow, and as a result had very low numbers of FC and *E. coli* entering the digestion process. Using quantitative PCR, the researchers suggested that nonculturable *E. coli* could exist in samples after digestion, and this nonculturable population was greater in thermophilic samples. After centrifugation, these nonculturable *E. coli* were "reactivated" or "resuscitated" which made them culturable. This would explain the immediate increase in *E. coli* or FC density measured after high solids centrifugation. Furthermore, after reactivation, the conditions in the cake were able to support additional growth of the bacteria, leading to an additional increase in the density of FC and *E. coli* during cake storage for several days.

Lebuhn et al. (2005) also reported that significantly higher densities of *E. coli* were enumerated using quantitative PCR compared to standard culturing methods for a process treating cattle manure that utilized anaerobic mesophilic, thermophilic, mesophilic reactors in series. Unfortunately, their study did not include dewatered samples to determine if SI and/or regrowth could occur. They also attributed the difference to the possible presence of undegraded DNA or nonculturable cells that are not enumerated by culturing techniques.

Jolis (2006) examined eight different liquid Class A biosolids from pilot and full-scale processes, and reported two of the eight samples exhibited growth during storage at 35°C. Storage of all samples at either 20 or 50°C did not show regrowth. It should be noted that the samples were not dewatered, which is believed to be the main cause of SI and regrowth, and instead, the analysis was performed on the liquid digested biosolids.

The body of knowledge to date is limited, but seems to indicate that for plants with the combination of digestion and centrifuge dewatering, there is a likelihood that SI and/or regrowth may be occurring. For utilities that utilize belt filter press dewatering, this probability appears to be lower. Significant additional research was identified to better understand this phenomenon, and the conditions under which it occurs. This information is needed to further our understanding of the sudden increase and regrowth in order to develop mitigation strategies. The specific objectives of this project are discussed in the next section.

1.2 Research Objectives

The second phase of the research was focused on better understanding the sudden increase and growth that occurs after digestion and dewatering, and, in particular, how process parameters impact this phenomenon of reactivation and regrowth. The main objectives of the research are summarized as follows:

- examine the impact of different process parameters such as reactor configuration, digestion temperature, solids reduction, solids retention times, dewatering equipment on the extent of the increases;
- determine if potentially pathogenic bacteria, such as *Salmonella*, also increased after dewatering of Class A and Class B biosolids.

- examine potential mechanisms to explain the sudden increase after dewatering followed by regrowth;

1.3 Nomenclature for Describing the Increases in Bacterial Densities

For the purpose of this report, it is important to note that we use the term "sudden increase" or "SI" to describe the increase in the culturable densities of bacteria measured immediately after dewatering compared to the culturable densities measured prior to dewatering. This term "sudden increase" is used a descriptor of the observation, and not to ascribe a mechanism to the increase. A detailed discussion of potential hypotheses and mechanisms to explain the sudden increase is reserved for Chapter 7.0. However, in several instances besides Chapter 7.0, the discussion in the report is relevant to the mechanisms, and in this case the term reactivation is used as a mechanistic descriptor of the phenomenon. In addition, the term "regrowth" is used to describe the increases in culturable densities measured during longer term storage after dewatering.

CHAPTER 2.0

RESEARCH APPROACH AND METHODS

2.1 Research Approach

To accomplish the objectives of this research, the project was divided into sub-components and an overview of these sub-components is provided in the following sections, with the detailed methodologies provided in subsequent sections.

2.1.1 Role of Reactor Hydraulics and Short Circuiting

Results from Phase I showed that single stage thermophilic plants were most likely to have greater increases after dewatering. The non-ideal hydraulics of these reactors often significantly reduce the actual detention time in the digester compared to the design values. In addition, by its very nature 50% of the fluid will have a detention time less than the design values. Interestingly, the one plant with completely stirred tank reactors (CSTRs) in series had no SI of *E. coli*. The Project Subcommittee (PSC) along with the research team hypothesized that the short-circuiting in typical plants may play a role in allowing inadequate treatment of the solids, and therefore inadequate pathogen destruction. Several full-scale plants have CSTR in series or batch digestion at thermophilic temperatures using the U.S. EPA time-temperature curve (U.S. EPA, 1999). These plants were sampled to examine the impact of reactor hydraulics, and reactors-in-series on SI potential. In addition, the one plant with thermophilic CSTRs in series that did not have SI, has the ability to dewater from each of the reactors in series, therefore, we performed a study to examine the SI and regrowth potential from each reactor in the series.

2.1.2 Examine Full-scale Process Parameters and How They Relate to the Destruction and/or the Potential for SI

During Phase I, three mesophilic, two thermophilic, and two temperature phased anaerobic digestion facilities were sampled. Based on this sampling, not enough data were collected to understand how process parameters may impact the destruction of indicator organisms, and under what conditions the bacteria enter a non-culturable state such that they can be reactivated. Instead, data were collected to support the hypothesis that bacteria were in a non-culturable state and could be reactivated during mechanical dewatering. Therefore, the Principle Investigators (PIs) proposed sampling additional digested biosolids from across North America to better understand how process parameters may relate to these phenomena. The data may also provide insight into systems that render the indicator organisms unable to be reactivated. This information will be potentially valuable to the field in terms of providing process modifications that can be used to overcome this problem.

2.1.3 Perform Lab Scale Time-Temperature Studies with the Inclusion of Quantifying Bacteria Using Quantitative PCR In Addition to Relying on Standard Culturing Methods

Design criteria for digesters were developed, in part, using time-temperature curves that related the time and temperature of digestion to the kill rates of certain pathogenic and indicator organisms. Unfortunately, these studies did not include analysis of non-culturable bacteria, and instead relied on standard culturing methods. As a result, these studies may have underestimated the actual densities of potentially viable indicator organisms. The project team in cooperation with other related on-going studies, performed several batch laboratory scale reactor studies to re-evaluate these curves, but also include the enumeration of *E. coli* using quantitative PCR. In addition, the potential for SI was examined using a lab scale process that simulates the shear in a high solids centrifuge. This simulated process has been developed for odor studies, and was investigated for use in SI studies.

2.1.4 Determine the Potential for Regrowth Of Indicator Organisms

One significant concern expressed by several utility managers involved in Phase I is the potential for regrowth of indicator organisms during storage. Results from Phase I demonstrated that FCs and *E. coli* grow quickly after SI and reach a peak concentration within 24-72 hours after centrifugation, after this time they begin to die-off slowly. This research performed several studies to examine the regrowth after SI, as well as the die-off of the indicator organisms after they reach the peak concentrations. The utility managers expressed great interest in knowing how they re-grow, and how fast they might reach peak concentrations, as well as methods to mitigate this problem. These issues will greatly impact their beneficial re-use programs.

This phase (Phase II) examined the impact of regrowth of bacteria. Presumably, the lack of growth or culturability in the digester sample suggests the growth conditions are not favorable for the bacteria. However, after SI, the bacteria seem to grow and thrive in the cake. This suggests conditions have changed which make for a more favorable medium for growth. For example, lysis or shear may increase the amount of growth-promoting substrate, resulting in regrowth if competition is not adequate. Conditions that promote this regrowth were examined, such as presence of growth-promoting substrate and absence of competition among organisms.

2.1.5 Determine if Potentially Pathogenic Bacteria Reactivate and/or Regrow After Dewatering

The increase of indicator bacterial densities after dewatering are a concern since these bacteria are suppose to be indicators of other potentially pathogenic organisms. Therefore, it is important to understand if pathogenic bacteria are reactivated or regrow after dewatering. It is not expected that viruses or helmith ova would reactivate or regrow since they need a host, and the focus is on bacterial pathogens. The key bacterial pathogen of concern is *Salmonella*, since it is a known pathogen, and it is the only bacterial pathogen with an U.S. EPA promulgated method for enumeration in biosolids. It is also part of the regulations for Class A biosolids. Therefore, a series of plants were sampled to determine if several potentially pathogenic bacteria, especially *Salmonella*, are reactivating and/or regrowing after dewatering.

2.1.6 Perform Mechanistic Studies to Better Understand the Mechanism by Which the Bacteria Are Reactivated and Regrow

It is hypothesized that bacteria enter the non-culturable state as a result of exposure to prolonged stress such as low nutrient and substrate conditions, unfavorable temperature, and/or low Fe content. During exposure to prolonged stress, bacteria change their metabolic activity, and also may change their physiology such as cell wall structure to survive the stress according to the VBNC concept (Baker et al., 1983; Tholozan et al., 1999; Signoretto et al., 2000; Heim et al., 2002). In the case of sub-lethal injury, the bacteria are injured by such stress, however they are not killed, and can repair given the right conditions. These bacteria survive the digestion process and are reactivated downstream by centrifuge dewatering. The conditions that support SI may include lysis and release of cell growth promoters, cationic polymer addition, and changes to the biosolids environment. Environmental changes may also support SI such as modification of pH, substrate availability, oxidation-reduction potential (ORP) and/or temperature. If conditions are favorable in the biosolids, regrowth of reactivated organisms may also occur and may be sustained during storage of biosolids.

Factors that affect SI (e.g. the release of growth promoting compounds) were investigated. One possible mechanism is that shear during high solids centrifugation causes some cell lysis, and this leads to the release of cytoplasmic autoinducers (AI). In addition, several experiments were performed in which free radical scavenging compounds were added to the growth media for FC and *E. coli* enumeration. These compounds, such as catalase and pyruvate, have been shown to increase the recovery of both VBNC and sub-lethally injured bacteria (McDonald et al., 1983; Martin et al., 1976; Mizunoe et al., 1999; Whitesides and Oliver, 1997; Reissbrodt et al.. 2000; Haro-Kudo, 2000; Grey and Steck, 2001; Reissbrodt et al., 2002; Gupte et al., 2003). For this research project, several studies were performed to examine the mechanisms by which the bacteria are reactivated. One study examined the use of media enhancements for the purpose of better recoveries. These enhancements included pyruvate, catalase, and various autoinducers. Another set of experiments was performed to examine the role of shear using a laboratory procedure to simulate high solids centrifugation. This procedure has already been developed and used for odor studies, but it has not been studied to see if SI of indicator bacteria also occurs. Therefore, preliminary studies were performed to determine if this method has the potential for causing SI and regrowth.

2.2 Methods

2.2.1 Sample Collection

Biosolids samples were collected aseptically in sterile containers. Samples generally included digester influent and effluent as well as the dewatered cake sample. It should be noted that the digester effluent was typically the feed to the high-solids centrifuge; however, many plants had storage prior to dewatering. In these cases, samples were also collected from storage. Samples to be analyzed using standard culturing techniques were immediately put on ice in a cooler, and transported by overnight shipping for testing to begin within 24 hours of collection as recommended by U.S. EPA (1999). The samples to be used for molecular methods were placed on dry ice and then shipped using overnight shipping. Prior to shipping of liquid samples for molecular analysis, the sample is centrifuged on-site at 14,000*xg* for 5 minutes. The supernatant is discarded, and the pellet placed on dry ice.

2.2.2 Enumeration of *E. coli* and Fecal Coliform

FC were enumerated using standard culturing methods (EPA, 2005). *E. coli* were enumerated using both quantitative PCR (qPCR) and standard culturing methods (see below for details). Compared to the first phase, the analytical method for enumerating *E. coli* was changed to real time polymerase chain reaction (RT-PCR) from competitive PCR (cPCR). This was done based in part on suggestions from reviewers. The advantages of RT-PCR compared to cPCR include faster analysis, greater sensitivity, and better reproducibility. The main disadvantage is the cost of the equipment to run RT-PCR. Through a separate funding program, Bucknell University purchased a RT-PCR unit on October 1st for use in this project. The research team developed a RT-PCR protocol including primers to enumerate *E. coli*, and the samples analyzed previously by cPCR were analyzed by RT-PCR for comparison. The designed primers were also compared side-by-side with primers developed by other researchers targeting various *E. coli* genes for accuracy and sensitivity check, which is described in later sections with details. The detailed RT-PCR methodology is provided in the following sections. For the purpose of the project, the term quantitative PCR or qPCR will be used to represent the enumeration of *E. coli* based on the number of *E. coli* DNA copies present in the sample.

2.2.2.1 Development of the Real-Time PCR Technique for Detection of *E. coli.*

The real-time PCR (RT-PCR) technique is a molecular technique that allows more consistent, sensitive, and easier quantification of target DNA than the competitive PCR technique used in previous reports (Bhat and Epelboym, 2004). It should be noted that the abbreviation RT-PCR is sometimes used in the literature to also refer to reverse transcriptase PCR and real-time PCR, but for the purpose of this report, RT-PCR is used only as an abbreviation for real-time PCR. The technique is based on the use of a real-time PCR system, which detects real-time fluorescence signals after every PCR amplification cycle (Higuchi et al., 1993). The concentration of unknown target DNA is then determined through amplification plot analysis along with an external known target DNA standard. The method was not chosen in prior reports due to the expense of the equipment; however, the Civil & Environmental Engineering Department at Bucknell University has recently provided funding to purchase this equipment. Therefore, this report provides information on developing and verifying the methodology with prior samples.

2.2.2.2 Primer Design for RT-PCR

Since it is recommended that the PCR product of a RT-PCR protocol be less than 500 base pairs (bps), a new set of PCR primers is necessary for the RT-PCR assay. The same *gadAB* gene was used as target for *E. coli* quantification, and primers were designed using the PrimerQuest^{SM} software provided by Integrated DNA Technologies. The new primer pairs contain a forward primer 5' - GCG TTG CGT AAA TAT GGT TGC CGA - 3' (gadrt-1), and a reverse primer 5' - CGT CAC AGG CTT CAA TCA TGC GTT - 3' (gadrt-2), which yield a final 305 bp PCR product. Both primers were analyzed through BLAST for their specificity to *E. coli.*

2.2.2.3 RT-PCR Analysis and Standard Curve Validation

The SYBR green based RT-PCR method was chosen for the quantification of *E. coli* with the Stratagene MX3005P Real-Time PCR system (La Jolla, CA). The SYBR Green I dye binds to the double strand DNA structure (PCR products) and emits fluorescence signals, which were measured at the end of each PCR cycle, thus allowing the monitoring of PCR products throughout DNA amplification. Each PCR reaction contains 12.5 µL of the Brilliant® SYBR® Green QPCR Master Mix from Stratagene (La Jolla, CA), 0.5 µM of each primer, 30 nM of reference dye (Rox), 10 µL of 10 ng sample DNA, and DNase/RNase free water to a final volume of 25 µL. The PCR program contains a 10-min initial denaturation at 95°C, followed by 38 cycles each of denaturation at 95°C for 30 sec, annealing at 57°C for 1 min, and extension at 72°C for 30 sec. The program also includes a final dissociation curve analysis for QA/QC of PCR products.

Serially diluted *E. coli* DNA ranging from two to 7620 copies was used as an external DNA standard for each RT-PCR analysis. Ten ng of *Psudomonas putida* DNA was also included in each standard serving as the background DNA. Duplicates or triplicates of each sample were run for each RT-PCR assay. An example of an amplification plot is shown in Figure 2-1. The horizontal line indicates the threshold cycles (Ct) of each plot for standard curve construction. The threshold cycle is defined as the cycle where fluorescence signal is significantly above the background, and is shown to be inversely proportional to the log quantity of the initial target amount. The standard curve for this experiment had an R-square of 0.988 indicating good consistencies and accuracy of the developed method (Figure 2-2). Figure 2-3 shows the final dissociation curve of the PCR products with only one single peak, indicating no non-target DNA was amplified. The concentrations of unknown DNA samples analyzed together with the standards were then determined using the standard curve equation (Figure 2-4). The improved method allows sensitive detection of a single copy of *E. coli* DNA in each reaction which is approximately 10,000 to 20,000 *E. coli* per gram dry solid.

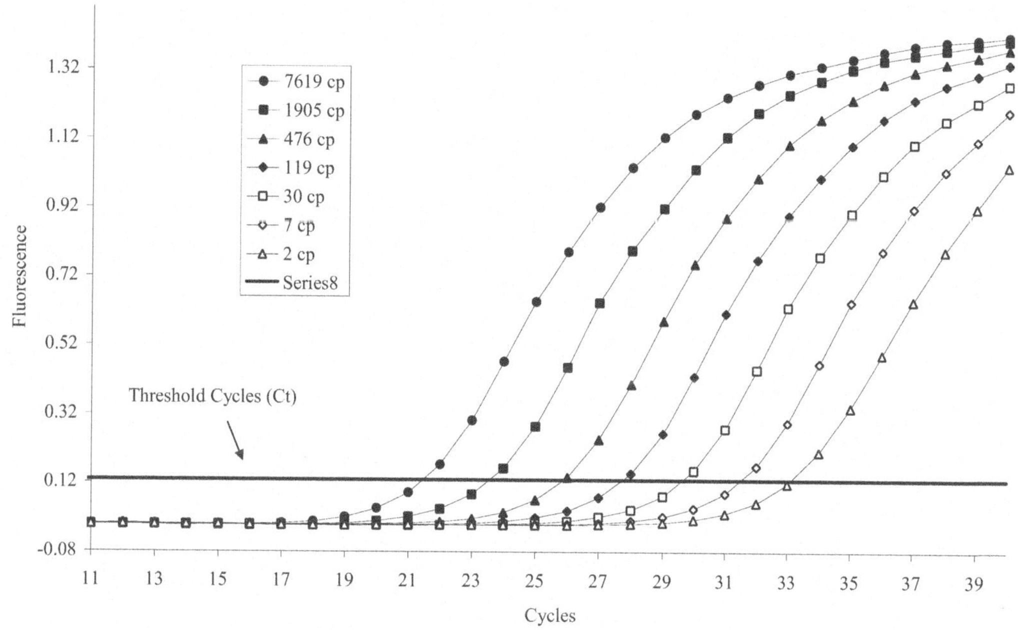

Figure 2-1. Amplification Plot for *E. coli* Standards.

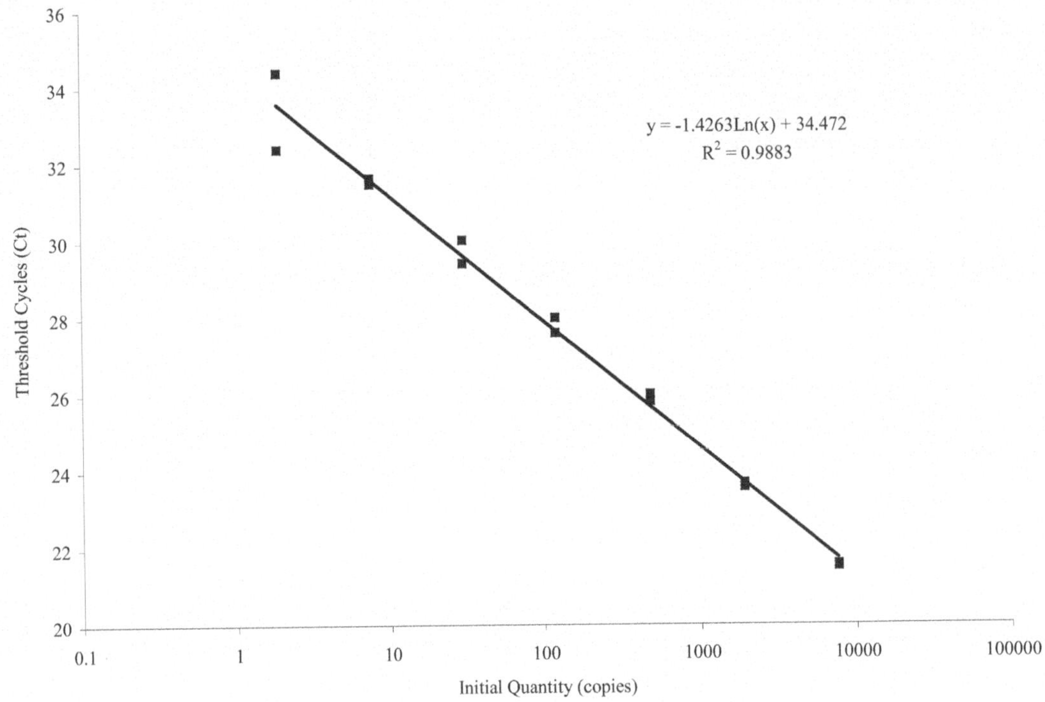

Figure 2-2. ***E. coli DNA*** **Standard Curve.**

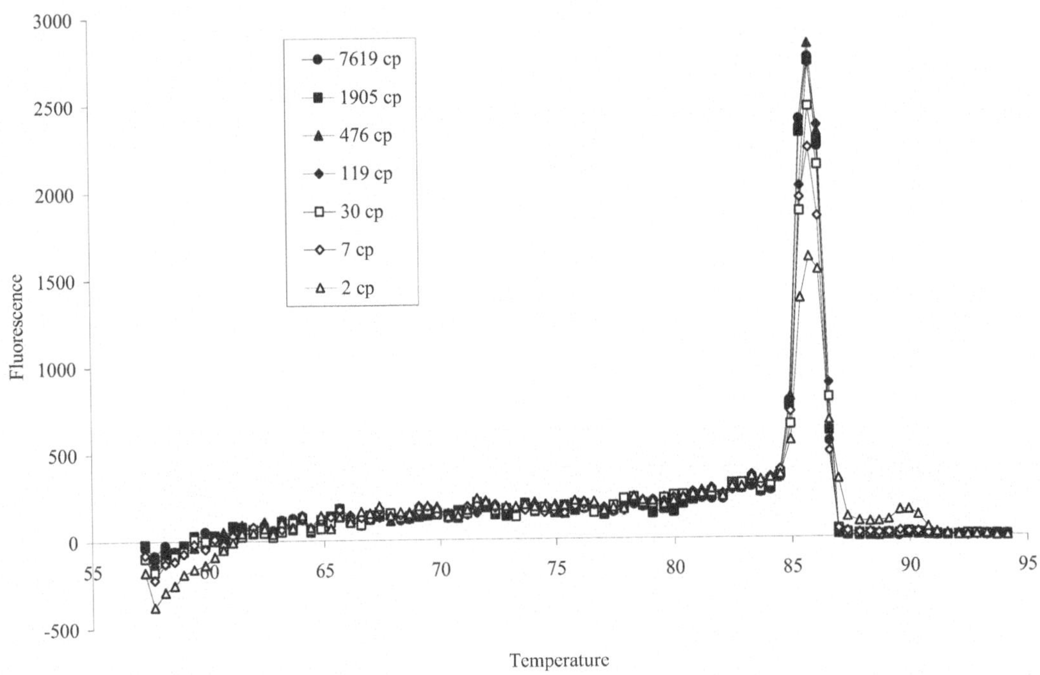

Figure 2-3. Dissociation Curve of the Final PCR Products.

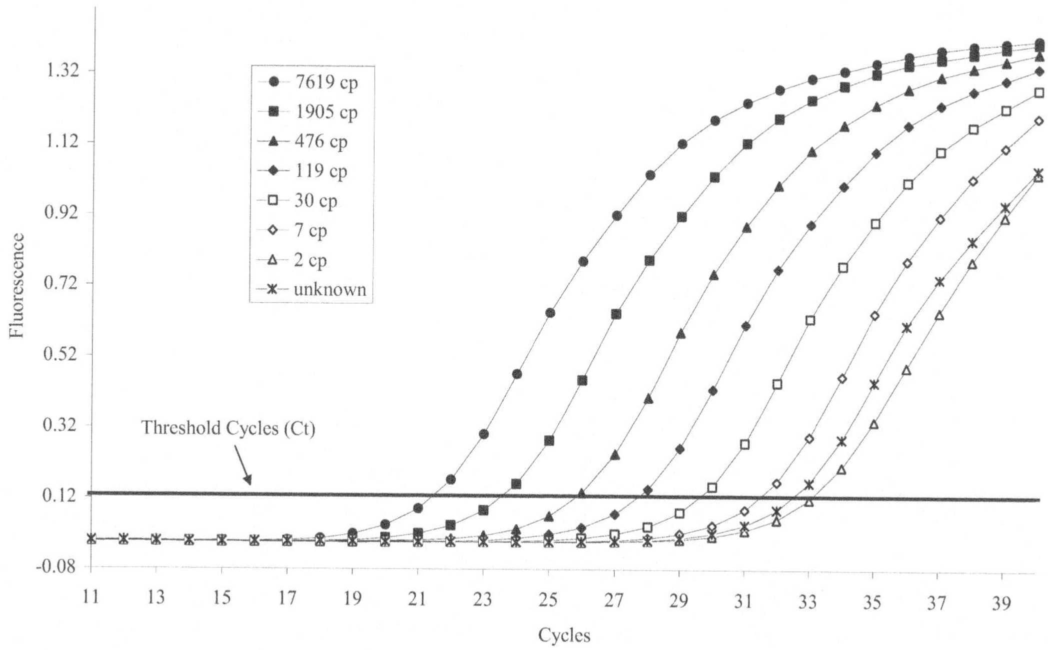

Figure 2-4. Amplification Plot of DNA Standards with Unknown Sample.

2.2.2.4 Comparison of RT-PCR and cPCR

Figures 2-5 and 2-6 compare the two quantitative PCR (qPCR) methods used in this research for Mesophilic Plant 1 and Thermophilic Plant 1. The results indicate the two different methods produced comparable *E. coli* concentrations. The scale of error bars indicate that RT-PCR and cPCR had similar variability overall. A key advantage of the RT-PCR method was it was able to enumerate *E. coli* in several of the samples for which the cPCR method could not. For example, in Mesophilic 1 the cPCR method did not measure any *E. coli* in the pumped cake sample, and in Thermophilic 1, the cPCR method did not measure any *E. coli* in the Digester 2 and the centrifuge cake samples. The detection limit was reduced by about 1-2 orders of magnitude with the use of RT-PCR compared to cPCR which is important for this study. Overall, the RT-PCR method had better detection performance than the cPCR method, which is consistent with prior reports (Bhat and Epelboym, 2004). It is important to note that since the cPCR method and RT-PCR methods agree quite well, the previous research based on cPCR results are valid. The main disadvantage was the higher detection limit associated with the cPCR method.

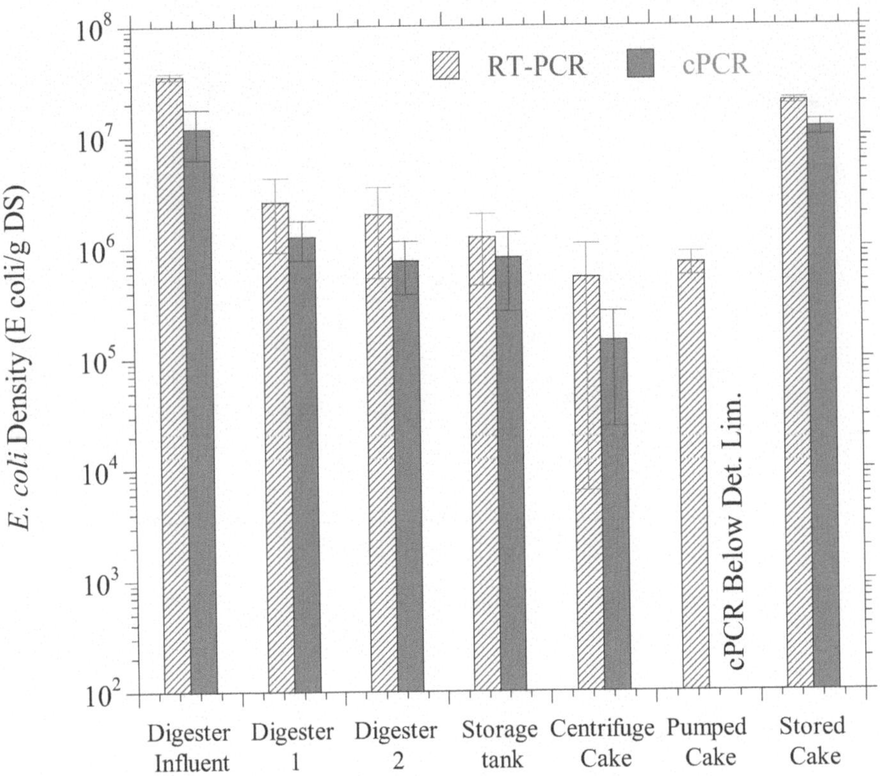

Figure 2-5. Comparison of RT-PCR and cPCR Results from Mesophilic Plant 1.

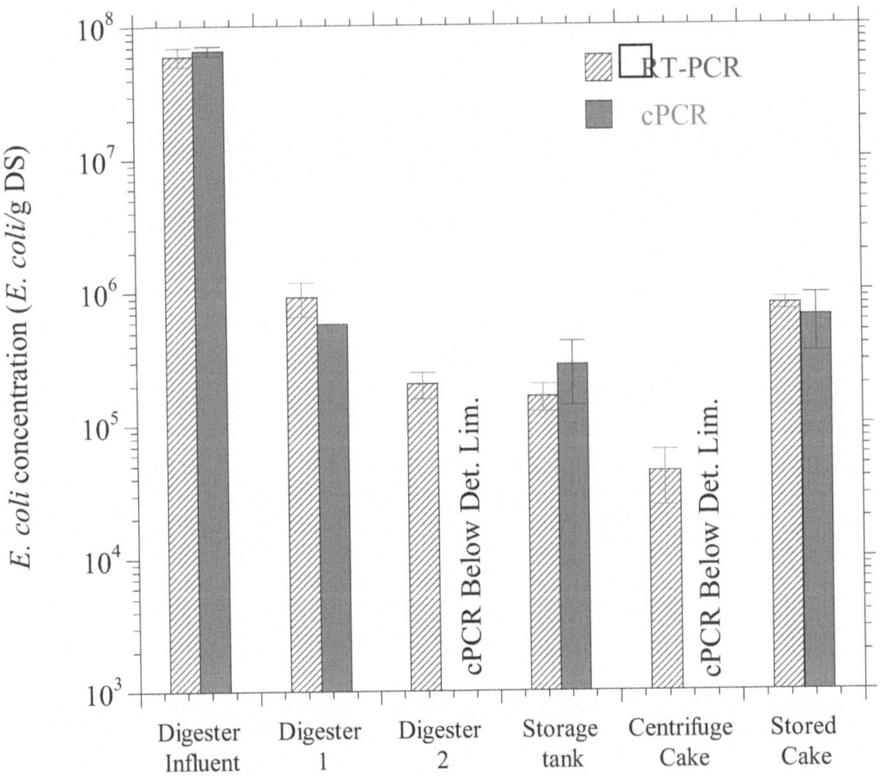

Figure 2-6. Comparison of RT-PCR and cPCR Results from Thermophilic Plant 1.

2.2.2.5 Comparison of Different RT-PCR Primers

In order to further validate the accuracy of the RT-PCR primers designed in this research and the validity of *gadA/B* gene as the target for *E. coli* enumeration, two additional RT-PCR primer sets, each targeting *E. coli uid* and *rRNA* genes, were tested side-by-side with the primers used in this research. Both primer sets were developed and verified by other researchers and PCR conditions were operated per the authors' recommendation (Table 2-1). The resulting standard curves of the three primer sets are parallel to each other, indicating that quantification of *E. coli* using these three primer sets can yield comparable results (Figure 2-7). Comparable *E. coli* enumerations of the field samples with three different *E. coli* genes further confirmed the validity of using *gadA/B* gene for *E. coli* quantification (Figure 2-8). It was suspected that positive enumeration of *E. coli* using molecular techniques from samples exhibiting no culturable *E. coli* might be an artifact of false PCR amplification of unknown bacteria in the sample. However, positive and consistent enumerations of *E. coli* with all three primer sets of a digested sludge with no-culturable *E. coli* (Figure 2-8, P-1 Digester Effluent), proved that it was indeed *E. coli*.

Based on the threshold cycle (Ct), it appears that RT-PCR targeting *rRNA* has the best sensitivity, followed by *gadA/B*, and *uid* (Figure 2-7). This result is likely due to more *rRNA* genes (7 copies) present in a single copy of *E. coli* chromosome than *gadA/B* (2 copies) and *uid* (1 copy). Nonetheless, both *rRNA* and *gadA/B* targeting primers were able to detect the lowest concentration of standard DNA, while *uid* targeting primers did not.

Table 2-1 Gene Targets and Primer Sequences for *E. coli* Enumeration.

Target gene	Primer Sequence	Annealing Temp.	Reference
gadA/B	F: 5' - GCG TTG CGT AAA TAT GGT TGC CGA - 3' R: 5' - CGT CAC AGG CTT CAA TCA TGC GTT - 3'	57°C	This research
uid	F: 5' - TTG CTG TGC CAG GCA GTT T - 3' R: 5' - ATC ATG GAA GTA AGA CTG C - 3'	55°C	Lleo et al., 2005
rRNA	F: 5' - GTT AAT ACC TTT GCT CAT TGA - 3' R: 5' - ACC AGG GTA TCT AAT CCT GTT - 3'	55°C	Malinen et al., 2003

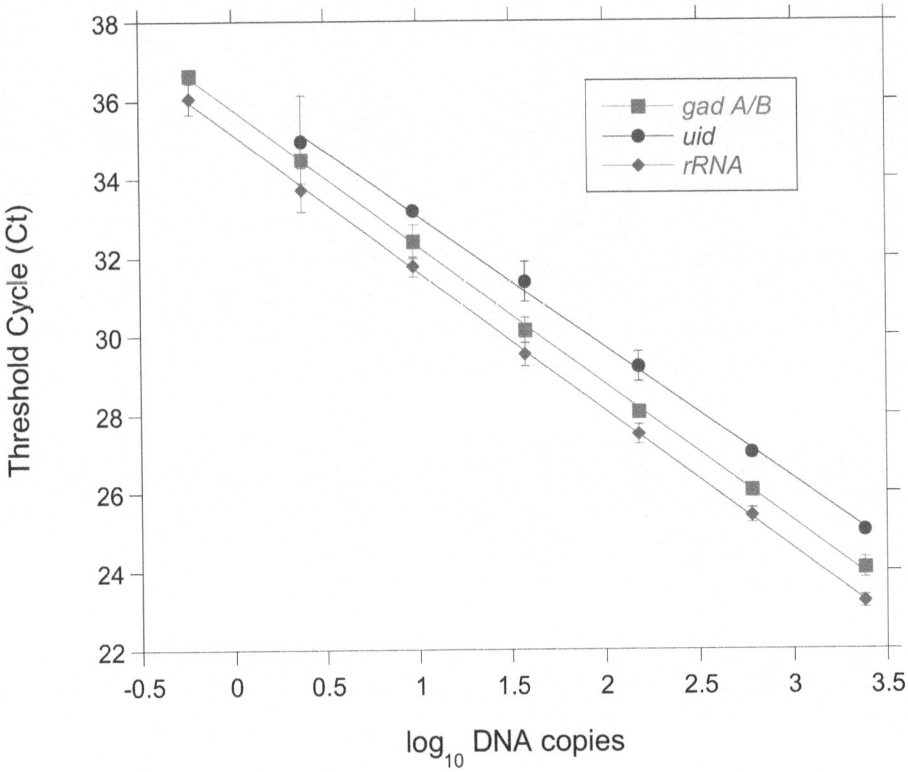

Figure 2-7. RT-PCR Standard Curves of Three Gene Targets.

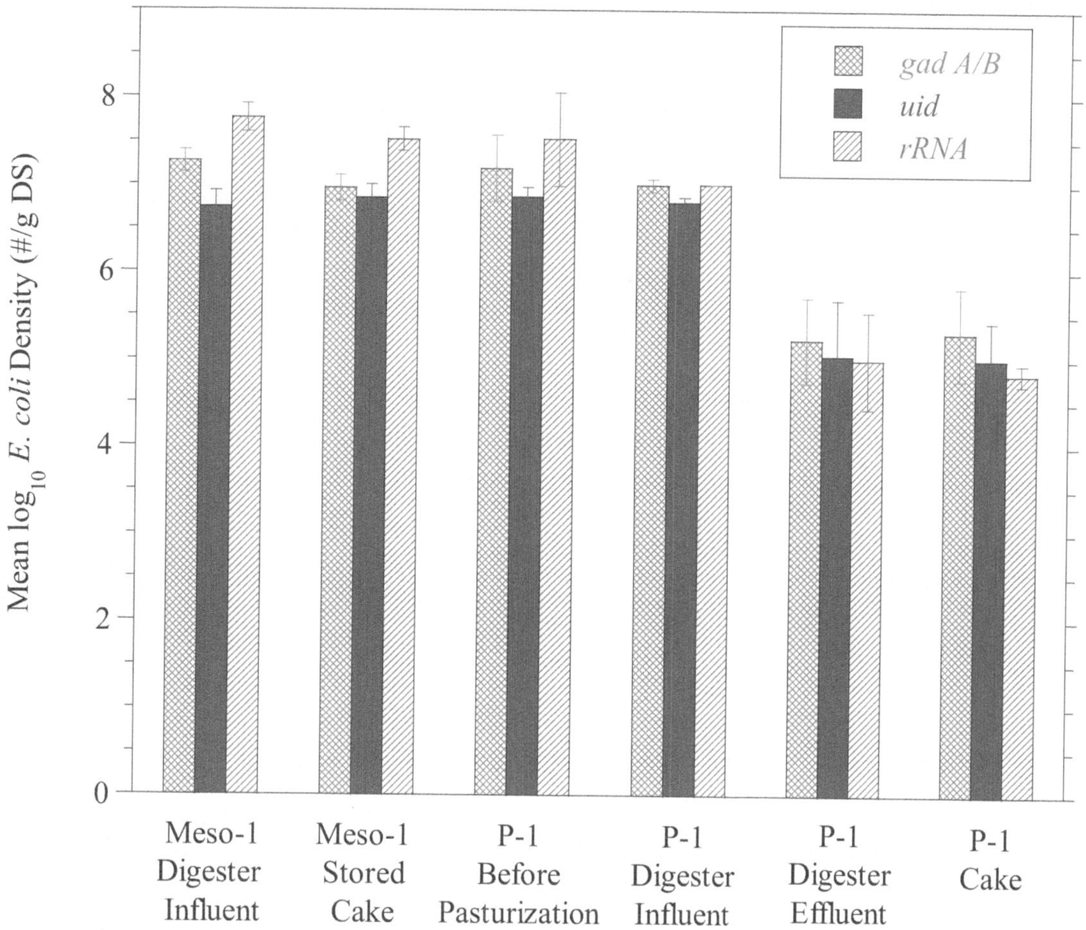

Figure 2-8. RT-PCR *E. coli* Enumerations with Three Gene Targets.

2.2.3 Enumeration of FC, *E. coli*, and HPC by Culturing Methods

The methods used for sampling and analysis of FCs were those promulgated by the U.S. EPA (1992, revised 1999 and 2005a) as required by the Part 503 regulations.

Sample processing. Samples were generally collected and processed in accordance with recommendations from U.S. EPA Method 1680 (U.S. EPA, 2005a). A Waring blender was used for processing the samples. The blender containers were 1.2 L stainless steel blender vessels with lids, and they were sterilized by autoclaving for 15 minutes at 121°C, 15 psi. Thirty grams of the sample were aseptically weighed and added to the blender containers containing 270 mL of sterile phosphate buffered water. Samples were blended for two minutes at 18,000 rpm before serial dilutions.

Fecal coliform analyses. The FC (MF) analyses were conducted according to SM9222D using m-FC media incubated at 44.5 ± 0.2°C for 24 hours. The FC MPN procedure (U.S. EPA Method 1680) was also used for enumerating fecal coliforms.

E. coli analysis. The *E. coli* MPNs were enumerated using the EC-MUG media following procedures described in SM9221F. The modified mTEC media was also used for *E. coli* MF enumeration following procedures described in U.S. EPA Method 1603 (U.S. EPA, 2005b).

2.3 Mesophilic Digestion SRT Experiment

The effect of anaerobic digester solids retention time (SRT) on FC and *E. coli* densities measured by SCM and/or qPCR was examined using laboratory scale, mesophilic anaerobic digesters that were operated by the Los Angeles County Sanitation Districts' (LACSD) Joint Water Pollution Control Plant (JWPCP) in Carson, California. The JWPCP is a large wastewater treatment plant that treats 350 mgd of wastewater and produces over 1650 wet tons/day of biosolids. The testing was conducted using five, 10-liter bench scale digesters that were fed with sludge produced from the JWPCP. Five different bench scale digesters with five different solids retention times were used during the testing. The SRTs that were examined included, 10, 15, 20, 30, and 40 days.

Sampling of the reactors occurred after steady state conditions were considered to be achieved for each of the units. Steady state for biological treatment systems usually requires two or more solids retention times, and the criteria used to assess steady state conditions were consistent gas production, low volatile acids concentrations and normal alkalinity concentrations. Personnel from the LACSD operated the bench scale digesters.

It should be noted that the operation of these reactors was performed as part of a separate WERF funded research project (03-CTS-9T - *Biosolids Processing Modifications for Cake Odor Reduction*). The PIs of this research are also involved in the WERF odor project, and were able to use the synergies of the two projects to perform this study. The PIs thank LACSD for sending samples for analysis of FC and *E. coli* during the digester operation.

The bench scale digesters were fed a mixture of thickened waste activated sludge (TWAS) and primary sedimentation tank sludge on a daily basis, using a draw and fill procedure. The two sludges were blended before addition to the bench-scale digesters at the same ratio that is used for the full-scale digesters at the JWPCP. The sludge mixture was 25% TWAS and 75% primary sedimentation tank sludge. The sludge volume added and removed was based upon maintaining the required SRT of each digester. After sludge removal and addition, each digester was placed upon a shaker table and the digester volume mixed for 15 minutes. Shaker table mixing has demonstrated in previous studies the ability to provide sludge mixing as effective as the JWPCP's full-scale digesters gas-recirculation systems. The bench-scale digesters were operated at mesophilic anaerobic digestion temperatures of 35°C by the use of heated water baths. The water baths were approximately 110 liter, insulated coolers in which two 10-liter digesters were placed. Two 1500 watt immersion heaters were used to heat the water baths and a programmable temperature controller regulated the temperature. Digester gas was collected into five separate 10-liter Tedlar bags. Daily gas volume was determined by emptying the Tedlar bags by the use of a calibrated one-liter jumbo syringe.

During the study the influent blend of feed sludge were analyzed for total and volatile solids, total COD and pH. The bench scale digester units were monitored for sludge temperature, weight and gas production. The digester gas was analyzed for methane, carbon dioxide, air and hydrogen sulfide. The sludge effluent of the digesters was examined for total and volatile solids, pH, total COD, volatile acids, alkalinity and ammonia. Digester content weight, temperature, gas volume and hydrogen sulfide concentrations were measured on a daily basis. Influent and effluent sludge analyses were performed two to three times each week. Digester gas constituent concentration analysis (except for hydrogen sulfide) was completed once each week. The analytical methods used for the bench scale digesters at the JWPCP are based upon procedures listed in *Standard Methods* or other long-established laboratory testing references, see Table 2-2.

Table 2-2. Methods Used In Testing of the Influence of Digester SRT on Odor Production.

Analysis	Los Angeles County Sanitation Districts Method Used for Analysis	Method Based Upon
Total Solids	LACSD Method 153	Based on *Standard Methods 18th Edition* – 2540B
Volatile Solids	LACSD Method 154	Based on EPA Method 160.4
Volatile Acids	LACSD Method 638	Based on *Standard Methods 18th Edition* – 5560C
Alkalinity	LACSD Method 305B	Based on *Standard Methods 18th Edition* – 2320B
Ammonia	LACSD Method 201	Based on *Standard Methods 20th Edition*– 4500 – NH_3B&E
Total COD	LACSD Method 403	Based on *Standard Methods 18th Edition* – 5220C
pH	LACSD Method 101	Based on *Standard Methods 20th Edition* – 4500 – H^+
Digester Gas – Methane, Carbon Dioxide and Air	LACSD Method 1203A&B	Supelco Inc. GC Bulletin 712F
Digester Gas – Hydrogen Sulfide	LACSD Method 281	Gastec Hydrogen Sulfide Tubes – 4H&M

2.4 Thermophilic Time-Temperature Experiments

A series of experiments were performed to examine the impact of time and temperature on FC and *E. coli* destruction. Laboratory scale digesters were operated by Chris Wilson and John Novak at Virginia Tech and the samples were analyzed using both SCM and qPCR. The reactor setup and operation is described below.

High-density polyethylene small batch fermentation reactors supplied by Hobby Beverage Equipment Company (Temecula, California) were chosen for this study. The conical bottom of these vessels was thought to be advantageous in terms of mixing and suspension of grit, similar to the full-scale application of egg-shaped anaerobic digesters. The nominal volume of each vessel was 6.5 gallons (25 liters) and was operated with an active volume of 22.5 liters. The reactor vessels were modified to accept a threaded stainless steel thermometer, also supplied by Hobby Beverage. Temperature control was achieved by circulating heated water through an external jacket (½" I.D. vinyl tubing wrap). Reactors were operated at 49°C, 51°C, 53°C, 55°C, and 57.5°C (± 0.2) throughout this study.

At startup, digesters were seeded with approximately 15 liters of mesophilic anaerobic digested sludge from Pepper's Ferry Regional Wastewater Treatment Facility (Radford, Virginia). Following initial seeding, observed digester gas production ceased within approximately 5 days and then gradually increased to a steady rate for each digester. This suggests that the mesophilic microbial community had been deactivated due to the increased temperature, and a lag phase was required for the thermophilic microbial community to grow within each reactor. Rebel™ wet-tip gas flow meters (Nashville, Tennessee) were used to monitor daily gas production rates by the anaerobic digesters. This startup regime helped to ensure that the microbial community within each anaerobic digester was unique, and solely dependant on the applied digestion temperature. After approximately 15 days of acclimation to the desired digestion temperature, daily feeding in the absence of wastage was used to bring the digester contents up to its final operating volume.

The digester feed consisted of a 1:1 ratio of primary and secondary solids from DCWASA Blue Plains Advanced Wastewater Treatment Facility (BPAWWTF) measured on a mass basis and diluted to approximately 3% total solids. In order to achieve the desired SRT, the digesters were daily batch fed 1.5 liters of blended sludge using a draw-fill procedure. Ice-packed primary and secondary solids were received weekly from BPAWWTF via overnight shipment. Solids were then blended at the appropriate ratio and homogenized using a 200-watt Hamilton Beach kitchen stick blender. Pre-screening of the feed sludge to remove large particles was not required.

Suspension and mixing of the reactor contents was achieved by digester gas recirculation. Variable speed (6-600 rpm control) peristaltic pumps manufactured by Cole Parmer (Vernon Hills, Illinois) and Cole Parmer Masterflex Tygon LFL-17 pump tubing were used for gas pumping. General gas recirculation throughout reactor operation was controlled at approximately 0.8 liters per minute (50% max.). Homogenization of the reactor contents prior to sampling was accomplished by increasing the recirculation rate to approximately 1.5 liters per minute (100% max.). A fixed solids mass balance across the digesters was performed and revealed greater than 99% recovery on average using this mixing regime. The increased recirculation rate was also applied for about 10 minutes after feeding to disperse the raw sludge throughout the digester.

2.5 Mechanisms of SI Experiments

2.5.1 Resuscitation with Enhanced Culturing Conditions

A number of experiments were performed to determine if recovery of nonculturable organisms could be increased with enhanced culturing conditions which has been observed by other researchers (Rockabrand et al., 1999; Makino et al., 2000; Reissbrodt et al., 2002; Martin et al., 1976; McDonald et al., 1983). These methods generally relied on addition of components to the media such as catalase, pyruvate, and α-ketoglutarate in order to scavenge oxygen radicals such as hydrogen peroxide (Mizunoe et al., 1999; Martin et al., 1976; McDonald et al., 1983). Ascorbic acid (vitamin C), a commonly used anti-oxidant, was also tested for its potential for media amendment. Similarly, autoinducers have been added to resuscitate VBNC bacteria (Reissbrodt et al., 2002) and norepinephrine was added to enhance production of autoinducers of bacterial growth (Lyte et al., 1996). For removal of inhibitors, several trials were performed that included a sludge washing step to remove potential inhibitors, and the addition of bentonite clay to adsorb potential inhibitors. The procedures used for each of these tests are described in the following sections.

2.5.1.1 Addition of Scavengers and Autoinducers

Several trials were performed in which scavengers of free radicals were added to the Lauryl Tryptose Broth (Difco) with 0.01 g/L Bromocresol Purple (Sigma). These additions were:

1. 1000 Units/tube Catalase (Sigma)
2. 1% Sodium pyruvate (Sigma)
3. 1% α-Ketoglutaric acid (Sigma)
4. 0.1% Ascorbic acid (Sigma)

Catalase Enhanced Media. Dilute 1000 U of catalase to total final volume of 1.0 mL with RO water. Typically prepare two (2) extra tubes worth of solution to account for filter loss. Filter sterilize solution using a 0.45 μm sterile filter apparatus (Nalgene). After stock presumptive media is autoclaved, aseptically transfer 1.0 mL of catalase solution to each tube. Refrigerate media after sterilization.

Sodium Pyruvate Enhanced Media. Prepare presumptive media with 1% sodium pyruvate. Adjust the pH of media to 6.8 ± 0.2 using 1.0 M HCl if necessary. Dispense 10 mL of media into culture tube. Autoclave media for 15 min at 240°F. Refrigerate media after sterilization.

α-Ketoglutaric Acid Enhanced Media. Prepare presumptive media with 1% α-Ketoglutaric acid. Adjust pH of media to 6.8 ± 0.2 using 10 M NaOH. Dispense 10 mL of media into culture tube. Autoclave media for 15 min at 240°F. Refrigerate media after sterilization.

Ascorbic Acid Enhanced Media. Prepare presumptive media with 0.1% ascorbic acid. An ascorbic acid stock solution (0.1 g/mL) was prepared and filter sterilized prior to usage. Exactly 0.1 mL of ascorbic acid solution was added to each sterile LTB test tube. Refrigerate media until before use.

Autoinducer and norepinephrine. Similarly, researchers have shown that the addition of autoinducers as well as norepinephrine can potentially resuscitate VBNC bacteria (Reissbrodt et al., 2002, Lyte et al., 1996). The following additives were amended to the media at the described concentrations:

- 50 μM 3-oxo-C8-HSL (autoinducer compound)
- 50 μM C6-HSL (autoinducer compound)
- 50 μM (-)-norepinephrine

2.5.1.2 Removal of Inhibitors

Several experiments were also performed to determine if compounds present in the sludges were inhibiting the growth of the indicator bacteria on the media. These experiments included cell washing and addition of bentonite.

Cell washing. Thirty mL of liquid sludge was placed in a sterile centrifuge tube and centrifuged at 14,000 xg for 5 min. The resulting pellet was washed twice with dilution buffer by re-suspending the pellet and then centrifuged at 14,000 xg for 5 min. The washed pellet was then brought up to 300 mL using dilution buffer, followed standard MPN sample preparation procedures.

Bentonite addition. The addition of bentonite clay to the media was tested as a possible adsorbent of inhibitors present in the sample. Bentonite was first autoclaved and exactly 5 grams of bentonite was added directly into blender and followed standard MPN sample preparation procedures. A different approach of using bentonite was also tested in order to provide consistent high dosage of bentonite in MPN tubes. In this method, exactly 5 grams of bentonite was mixed with 300 mL LTB media, dispensed into individual MPN tube and sterilized through autoclaving.

2.5.2 SI by Simulating the Shear of a High Solids Centrifuge

It is believed that the shear, or uneven forces applied to the solids during full-scale centrifuge dewatering may contribute to SI and regrowth. Similar research on odors from biosolids has shown that shear is important to odors, and that centrifuge dewatered cake has a much higher odor than lower shear belt filter presses (Murthy et al., 2003). Therefore, simulating a field centrifuge in the lab may provide a means to study SI and regrowth without the need of a full-scale centrifuge. Research at Bucknell University that was funded by WERF (Project 01-CTS-1) as well as the Mid-Atlantic Biosolids Association (MABA) has examined the impact of both shear during conditioning and shear during dewatering on the production of odor causing compounds from the cake. A protocol for conditioning, dewatering, and processing solids has been developed that generates a cake with a similar odor production profile as a high solids centrifuge. This process is termed the high solids centrifuge simulation process (HSCSP). This method may also cause SI and regrowth which would provide a valuable tool for research. Therefore, several experiments were performed to determine if this method could cause SI and regrowth similar to a field centrifuge. The method is described in detail below.

Conditioning. First, the sludge is conditioned using cationic polymer. The polymer used is typically provided by the plant for which the study is being performed. In other cases, stock polymer from Polydyne or Stockhausen are used which are available to Bucknell University. One liter of solids is placed in a baffled reactor, and polymer is added to the solution. The sludge is mixed at a predefined mixing intensity and time, to achieve a given energy input or shear, which is measured by the Gt value. G is known as the velocity gradient, and is a measure of mixing intensity with units of $time^{-1}$. The symbol, t, is time and the product of the two, Gt is a unitless parameter used to quantify the overall energy imparted to the solution during mixing. The typical Gt value used for these experiments is 100,000. After mixing, the capillary suction time (CST) is measured. This process is repeated with varying polymer dosages until the optimum polymer dose is obtained, as measured by the polymer dose that produces the minimum CST.

Dewatering. The sludge conditioned with the optimum polymer dose is then placed in 250 mL Nalgene bottles and dewatered using a laboratory centrifuge at 3000 x g, for 10 minutes. After centrifuging, the supernatant is removed (this centrate can be analyzed for different constituents as well), and the cake is combined from the different tubes. Typically, about 200-300 g of wet cake are generated, with solids contents between 17-20%. This cake is then pressed using a laboratory scale belt filter press, to achieve cake solids in the range of 25-30% which is typical for full-scale centrifuges.

Processing. Next, the wet cake samples are processed to simulate the high shear experienced in the full scale centrifuge. To do this, the cake is processed through a meat-grinder apparatus that pushes the cake forward using a scroll-conveyor, followed by extrusion through a small opening at the end of the conveyor. This meat grinder is an attachment on a KitchenAid mixer.

After processing the cake samples were analyzed for FC and *E. coli* as described previously. All equipment was sterilized by autoclaving or boiling for 15 minutes prior to use in the HSCS process.

2.6 Quality Assurance and Statistical Analysis

2.6.1 Sample Collection, Handling, and Analysis

A number of processes and approaches have been used to assure collection of quality data throughout this project. This includes sample collection, sample analysis, and data analysis. A summary of these measures are provided in the following sections

Prior to sampling at a field site, a sampling protocol was developed in coordination with the field site. During the sampling event, U.S. EPA recommended protocols were followed for sampling, handling and storing (U.S. EPA, 1999). In addition, for every sampling event the same person who is knowledgeable with the appropriate protocols (Dr. Yen-Chih Chen) collected and processed the samples. For samples to be analyzed by culturing methods, the samples were immediately placed in a cooler, on ice, and shipped via overnight courier to the respective laboratory to remain within the recommended holding times. The majority of the field samples were sent to Hoosier Microbiological Laboratory (HML). HML conforms to ISO 17025 standards and is NELAP-certified. NELAP is a national accreditation for environmental laboratories that have to identify two performance evaluations (PEs) per year correctly, maintain extremely stringent QA/QC controls, and have one onsite evaluation every 2 years. HML is the only lab to have successfully isolated and identified *E. coli* Type 0157:H7 from biosolids. It also is the only lab in the United States with full NELAP accreditation for testing for Class A pathogens from wastewater solids. All field samples are analyzed, at a minimum, in triplicate to provide enough data for statistical analysis. The DNA extraction and qPCR analysis are performed by an expert in molecular methods, Dr. Yen-Chih Chen, who has significant experience with biosolids samples, and the application of molecular methods for evaluation of water and wastewater samples. In addition, the target DNA sequence used for *E. coli* analysis has been compared to other gene targets, with very good comparisons being found as discussed in Section 2.2.2.5, supporting the validity and specificity of the primer set for *E. coli*. In addition, replicates and standards are used with the molecular methods, and the PCR products are analyzed for continuity as additional quality assurance methods

2.6.2 Statistical Analysis of SCM and DNA Methods with Replication

Triplicate analyses of samples were performed in all field sampling and most laboratory tests except for a few laboratory MPN tests. Triplicate analyses are typically used to account for sample variation especially for those collected in the field. Some laboratory tests, with proper mixing prior to sampling, were performed without replication (only on MPN tests) due to the high number of samples that needed to be processed which imposed time limitations on the number of replicates that could be performed. Statistical analysis is still feasible for these tests because of the replications built within the MPN method.

For samples with replication, the mean was calculated using the geometric mean. For statistical comparisons between samples, $\log_{10}$ transformation was first performed, and the one-way ANOVA (F statistics) was used to evaluate differences among all treatments. Further pair-wise comparisons were performed to determine significant difference between paired treatments if desired, using either the Least Significant Difference (LSD) or Tukey's method.

2.6.3 Statistical Analysis of SCM (MPN) without Replication

The MPN test uses multiple or replicate tubes for each dilution which, through appropriate statistical analysis, provides an estimate of microbial densities and confidence intervals, but it does not allow users to compare sample means between different treatments based on this confidence interval. However, since the MPN procedure has "internal" replication (or multiple) tubes for each of the dilution series of the test, certain statistics can still be performed for treatment comparisons (Haas et al., 1999). The method is based on a null hypothesis that the MPN result of the dilution experiment for two samples are fit by a common mean. A likelihood ratio is then constructed and tested against the chi-squared distribution with a critical value of 0.05. Examination of Poisson distribution or goodness of fit of each set of MPN measurement is also tested to assure consistency between dilution and within the sample. We have created a spreadsheet to perform this calculation, and the spreadsheet is available (with instructions) upon request. The equations and procedures are summarized by Haas et al. (1999).

CHAPTER 3.0

FIELD SAMPLING AND EFFECTS OF DIGESTION AND DEWATERING: RESULTS AND DISCUSSION

3.1 Results of Field Sampling Events

Nine, full-scale plants were sampled as part of this research effort. These included three thermophilic facilities, a plant with pre-pasteurization followed by mesophilic digestion, and five conventional mesophilic facilities. The thermophilic plants provided a unique opportunity to study the impact of reactor hydraulics on FC and *E. coli* densities in the system. The Thermophilic Plant 1 has two completely stirred tank reactors (CSTRs) in parallel whereas Thermophilic Plant 2 has four CSTRs in series. Since reactors in series approach plug-flow, this process configuration will reduce short-circuiting associated with a single stage CSTR. In addition, the plant with pre-pasteurization uses a batch process to achieve Class A time-temperature requirements.

The mesophilic facilites also provided some different processes for comparison. One of the mesophilic plants had single stage CSTRs and the other plant had two CSTRs in series for comparison. In addition, the five mesophilic plants included two with centrifuge dewatering and three with belt filter press (BFP) dewatering. This allowed comparison of the effect of dewatering equipment on SI and/or regrowth. Similarly, one plant performed a rotary press trial. During the different sampling events, not all samples from each location in the plant were taken, and sample locations depended on the purpose of the event.

3.1.1 Thermophilic Plant 1

This facility utilizes anaerobic, thermophilic digestion with two CSTRs in parallel. A summary of relevant process parameters for the month prior to sampling are shown in Table 3-1. The effluents from the digesters flow into a storage tank and then are dewatered by a Sharples high solids centrifuge, Model DS 705. A cationic polymer is used for dewatering with a typical dose of about 4.4 kg/tonne DS. The mean cake solids for the month were 34.5%.

The density of FCs from each of the digestion and dewatering processes is shown in Figure 3-1. The influent to the digesters had 10^7 MPN/g DS which is within the typical range for most plants. The first thermophilic digester had an effluent concentration of approximately 67,000 MPN/g DS, and the second digester was less than 20 MPN/g DS. In Figure 3-1, the second digester effluent was assigned a value of 10 MPN/g DS simply for graphing purposes, although it is recognized that this is likely not the actual density. During storage prior to dewatering, the two digester effluents are mixed, and stored for a period of time. After storage, the FC density was approximately 200 MPN/g DS. Cake samples collected immediately after centrifugation had an average FC of about 15,000 MPN/g DS, an increase of about two orders of magnitude. After several days of storage in the biosolids storage hopper at the site, the FC concentration increased by two more orders of magnitude to approximately 1.6×10^6 MPN/g DS. These results demonstrate the pattern found in the Phase I research where relatively low FC densities can be measured after digestion just prior to dewatering, while immediately after dewatering, high densities are measured, followed by growth during storage of the cake.

Table 3-1. Summary of Digester Operational Parameter for Thermophilic Plant 1.

Parameter	Digester 1	Digester 2
Primary/Secondary Ratio	100% Primary	100% Primary
Digester SRT	15-20 d	Unknown[1]
Digester Temperature	54.8°C	55.3°C
Mixing	Gas mix with bottom diffusers	Gas mix with bottom diffusers
Digester pH	7.1	7.3
Digester Alkalinity	3599 mg/L as $CaCO_3$	4012 mg/L as $CaCO_3$
Digester Volatile Acids	521 mg/L	457 mg/L
Total Solids	2.6%	1.5%
Volatile Solids	76%	73%
Volatile Solids Destruction	64%	15%

[1]See text below for description.

A similar pattern was observed for *E. coli* densities measured by SCMs as shown in Figure 3-2. Both MPNs and membrane filtration (MF) techniques were used for the analysis, and depending on the dilution, the appropriate measurement is shown. The mean *E. coli* densities were approximately 2.5×10^7 cfu/g DS in the digester influent. The effluent from Digester 1 was approximately 10^4 cfu/g DS, and the effluent from Digester 2 was less than 10 MPN/g DS. At the time of sampling Digester 2 was experiencing atypical operation due to plant maintenance as solids that were being cleaned from the storage tank were being fed to Digester 2. These solids were very low in FC, and likely skewed the results. In comparison, Digester 1 was operating under normal conditions. During storage, the *E. coli* density increased to about 500 after mixing and storing the two sources, and the cake samples had 18,000 MPN/g DS.

For comparison, the RT-PCR results for *E. coli* enumeration are also plotted in Figure 3-2. Similar to most plants, the RT-PCR and SCMs for *E. coli* produced similar results in the digester influent. From Phase I research, it was found that the bacteria in digester influent were not typically found to be in the non-culturable state, and therefore, the SCMs and RT-PCR enumerations were typically in good agreement. However, as digestion proceeds, more bacteria enter the non-culturable state. The difference between the mean *E. coli* densities measured by RT-PCR and SCM in the effluent from Digester 1 is about one order of magnitude. Digester 2 had a slightly lower *E. coli* density measured by RT-PCR, however, the *E. coli* density measured by SCMs were <1 MPN/g DS. In the storage tank after mixing of the two digester effluents, the *E. coli* density measured by RT-PCR was about three orders of magnitude greater than the SCM density. Immediately after dewatering, the SCM density increased by about two orders of magnitude, and the density matched well with the RT-PCR value (the RT-PCR value decreased by almost an order of magnitude after dewatering). The immediate increase in SCM measured density is suggestive of a transition from a non-culturable to a culturable state as a result of the dewatering process or better enumeration. Cake storage resulted in an additional increase in the densities, and the RT-PCR density in the storage tank corresponded well with the FC density (no *E. coli* density was measured in the stored cake due to insufficient dilutions that were used).

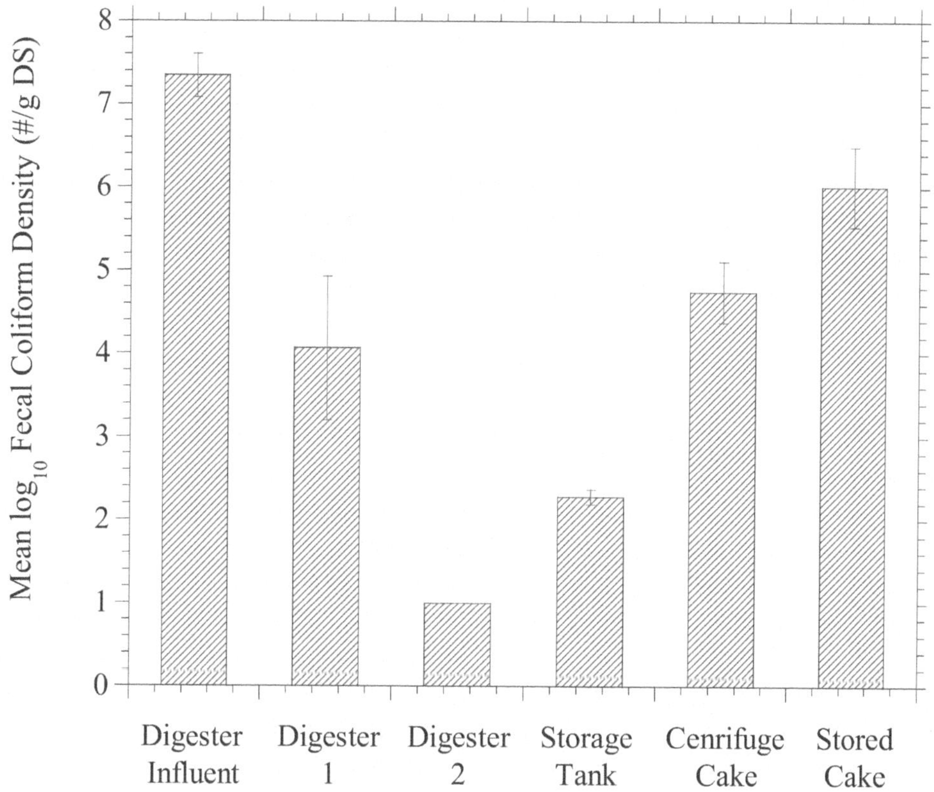

Figure 3-1. FC Density by SCM Through the Thermophilic 1 Digestion and Dewatering Process. (Error Bars Represent One Standard Deviation).

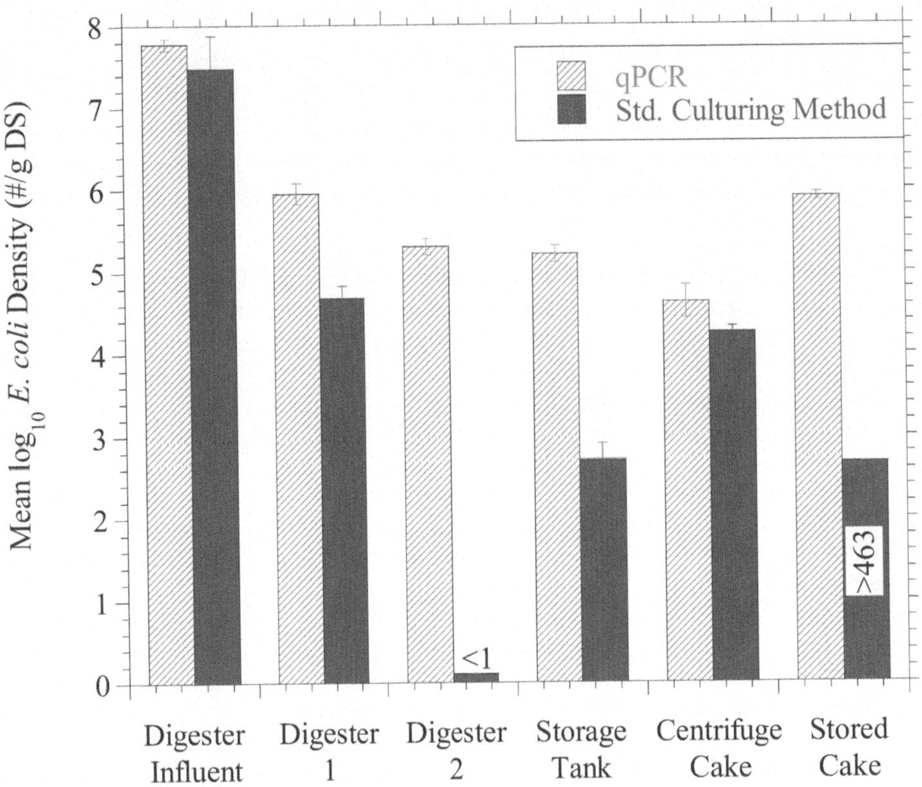

Figure 3-2. ***E. coli*** **Density Through the Thermophilic Plant 1 Digestion and Dewatering Process as Measured by SCMs and RT-PCR. (Error Bars Represent One Standard Deviation).**

The total DNA extracted after each process is plotted in Figure 3-3. It is interesting to observe that after the first stage of digestion, the extracted DNA steadily declines through each process. This indicates that the total mass of viable microorganisms is gradually declining through the treatment process, even during storage when FC and *E. coli* are increasing.

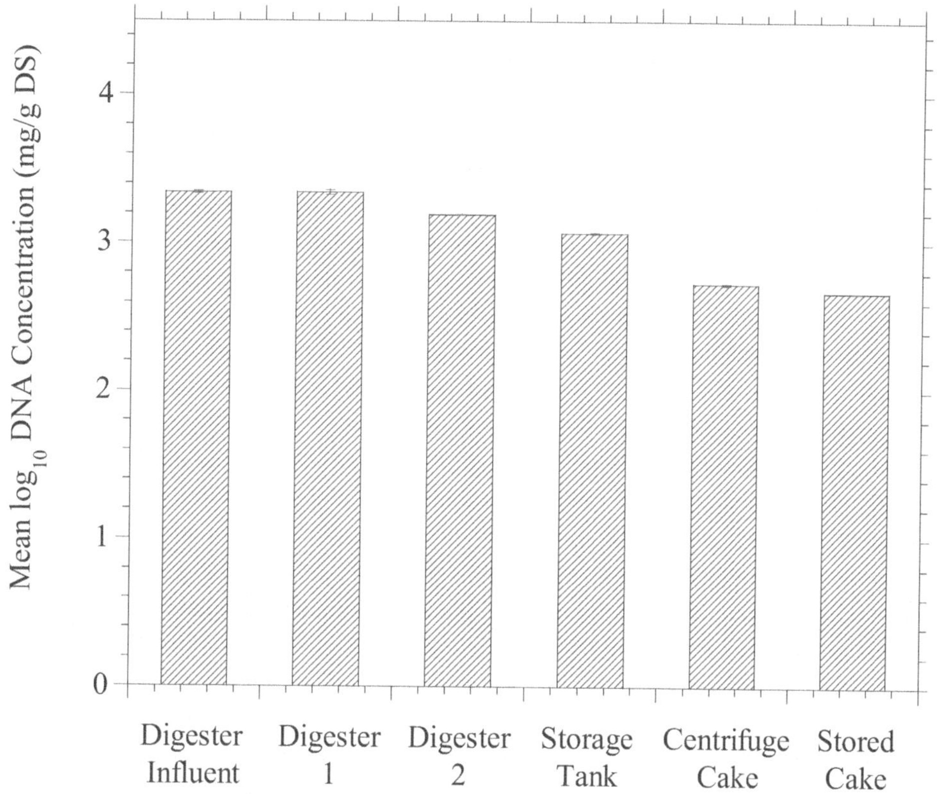

Figure 3-3. DNA Concentration Measured in the Thermophilic Plant 1 Digestion and Dewatering Process. (Error Bars Represent One Standard Deviation)

3.1.2 Thermophilic Plant 2

During the sampling event, Thermophilic Plant 2 was utilizing a three stage digestion process, with reactors in series, and a storage tank prior to digestion, effectively providing a fourth stage in the process. A summary of the digestion process parameters of interest are provided in Table 3-2. This plant was sampled during the Phase I research, but only the influent to the first stage and effluent from the last stage was sampled. During this sampling event, samples were collected from each of the stages as well as the storage tanks prior to dewatering. This was done to determine how multi-stage digestion may impact the density of FC and *E. coli*. The effluent from the storage tank is dewatered on Sharples high solids centrifuges Model DSX 706, with a mean cake solids of 30.2% during the month when the sampling event occurred. A cationic polymer was used with a dosage of 7.2 kg/tonne DS.

Table 3-2. Summary of Digester Operational Parameter for Thermophilic Plant 2.

Parameter	Stage 1	Stage 2	Stage 3	Storage
Primary/Secondary Ratio				
Digester SRT	16-19 d	≈1-2 d	≈1-2 d	≈1 d
Digester Temperature	55.3	53.8	54.4	51.7
Mixing	Gas – draft tubes	Gas- bottom diffusers	Gas- bottom diffusers	Gas- bottom diffusers
Digester pH	7.6	ND[1]	ND	ND
Digester Alkalinity	5228 mg/L	ND	ND	ND
Digester Volatile Acids	615	ND	ND	ND
Total Solids	2.1%	ND	ND	ND
Volatile Solids	70%	ND	ND	ND
Volatile Solids Destruction	60%	ND	ND	ND

[1]ND=no data available

The FC results are shown in Figure 3-4. Similar to the Phase I research, no SI occurred after dewatering for this plant. This plant has shown consistent inactivation of the FC organisms, at least to the point where SI and regrowth do not occur. The first stage substantially reduced the numbers of culturable organisms. The concentrations of *E. coli* in each process measured by RT-PCR and the SCM method are shown in Figure 3-5. The *E. coli* density in the influent measured by both methods was very similar. The RT-PCR method enumerated approximately $2x10^4$ *E. coli*/g DS after the first stage, while the SCM did not detect any *E. coli*. This is similar to Thermo-1, in that after single stage thermophilic digestion, RT-PCR measured about $9x10^5$ *E. coli* cells/g DS and had a slightly shorter SRT. The SRT at Thermophilic Plant 2 for the first stage was between 16-19 days. After the second stage which had and SRT of 1-2 days, the *E. coli* measured by RT-PCR was below the detection limit of approximately 10^4 *E. coli*/g DS, and similar results were found for the third stage, storage tank, cake and pumped cake.

The results suggest that the multi-stage process was capable of destroying the indicator organisms such that SI did not occur. As discussed in the next section (Section 3.2), regrowth during cake storage also did not occur. It is interesting that the first stage effluent from Thermophilic Plant 2 had similar *E. coli* densities measured by RT-PCR to the single stage Thermophilic Plant 1. Both of these single stages had similar SRTs. The additional SRT of the other stages in Thermophilic Plant 2 as well as the reactors-in-series appears to be important for destruction of indicator organisms. However, it is likely the reactor-in-series configuration provides the most benefit, since the additional SRT is only 1-2 days in the second reactor. Therefore, the reactors-in-series would help reduce short circuiting which might make the difference in reducing the densities of the indicator bacteria.

An interesting question that remains is would there still be possibility of SI and/or regrowth after the first or second stage? Additional testing was performed at this facility to examine if SI and/or regrowth could occur from Stage 1 and Stage 2 effluent, and the results are discussed in Section 3.1.3.

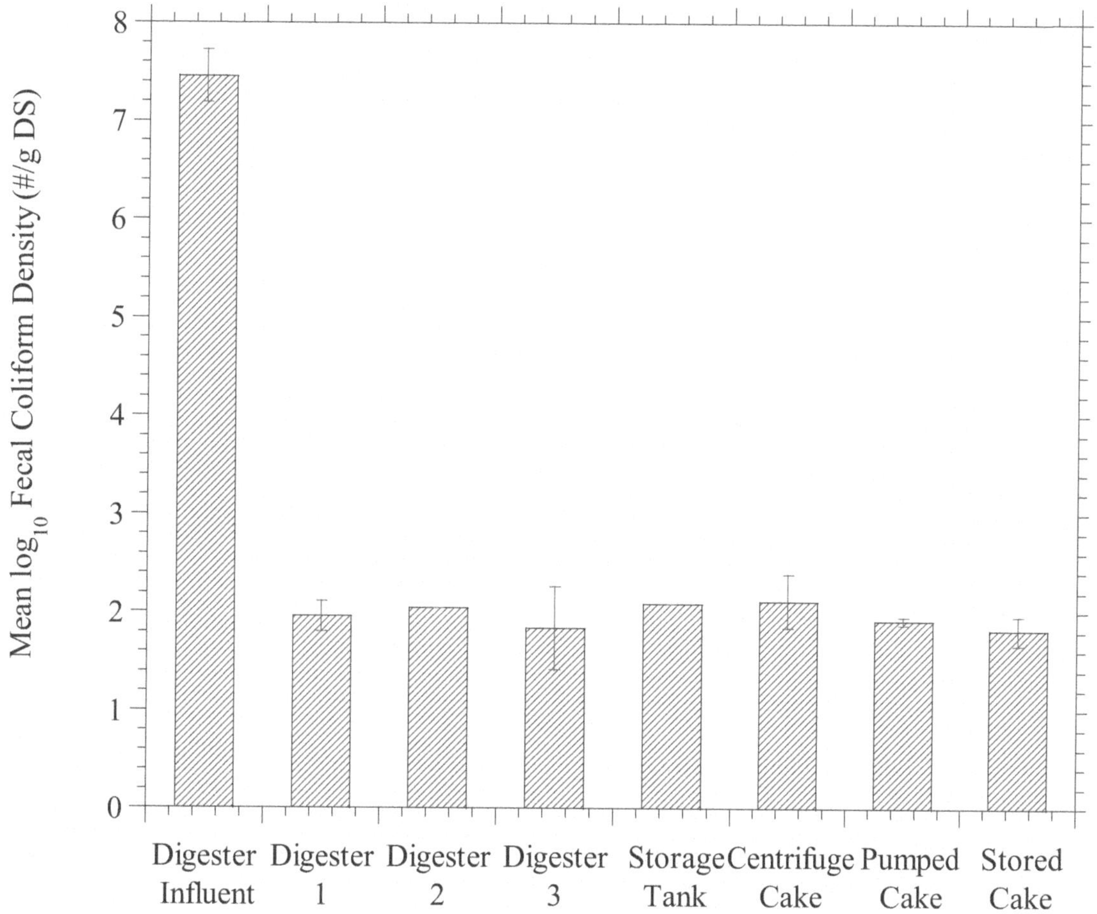

Figure 3-4. FC Densities Through the Thermophilic 2 Digestion and Dewatering Processes. (Error Bars Represent One Standard Deviation).

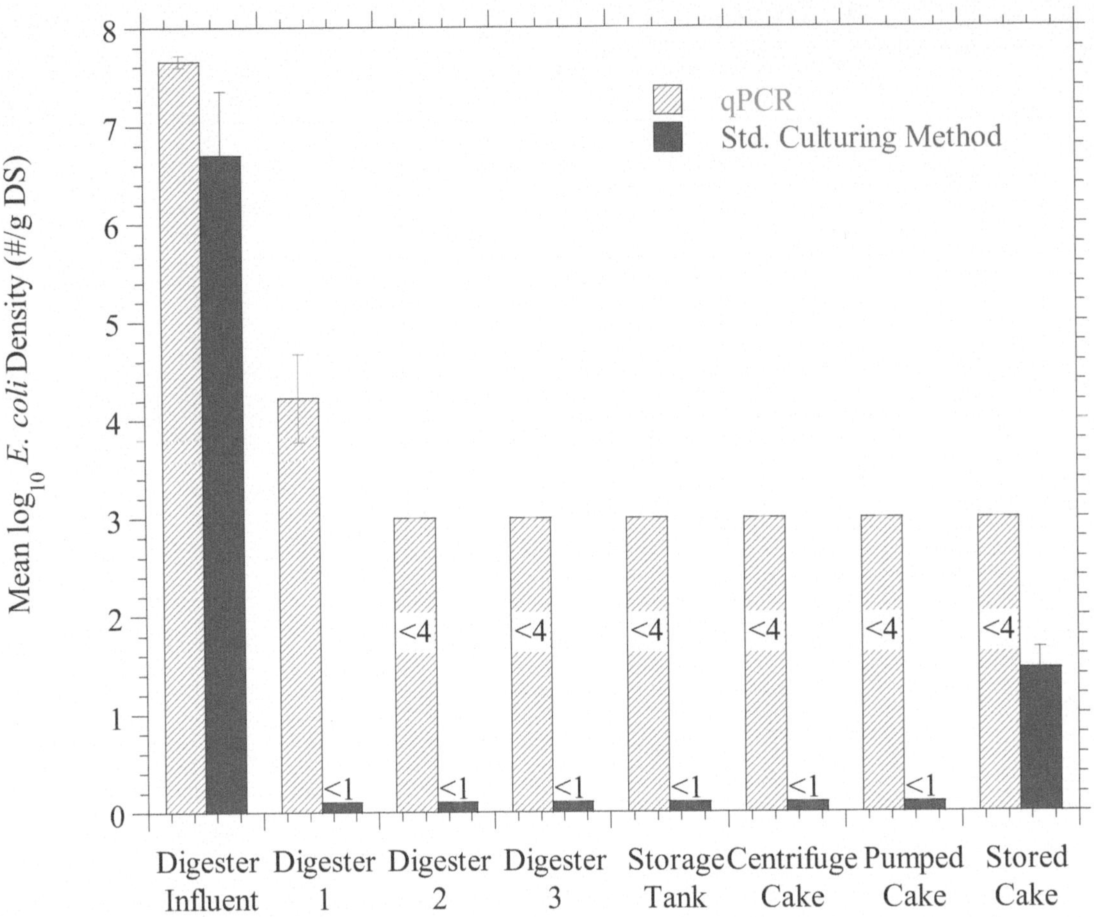

Figure 3-5. *E. coli* Density Through the Thermophilic 2 Digestion and Dewatering Process as Measured by Culturing Methods and RT-PCR. (Error Bars Represent One Standard Deviation).

The total DNA concentration extracted after the different process stages is shown in Figure 3-6. The three stage digestion process decreased the total DNA concentration by about 65% compared to an average of 15% for Thermophilic 1 digestion.

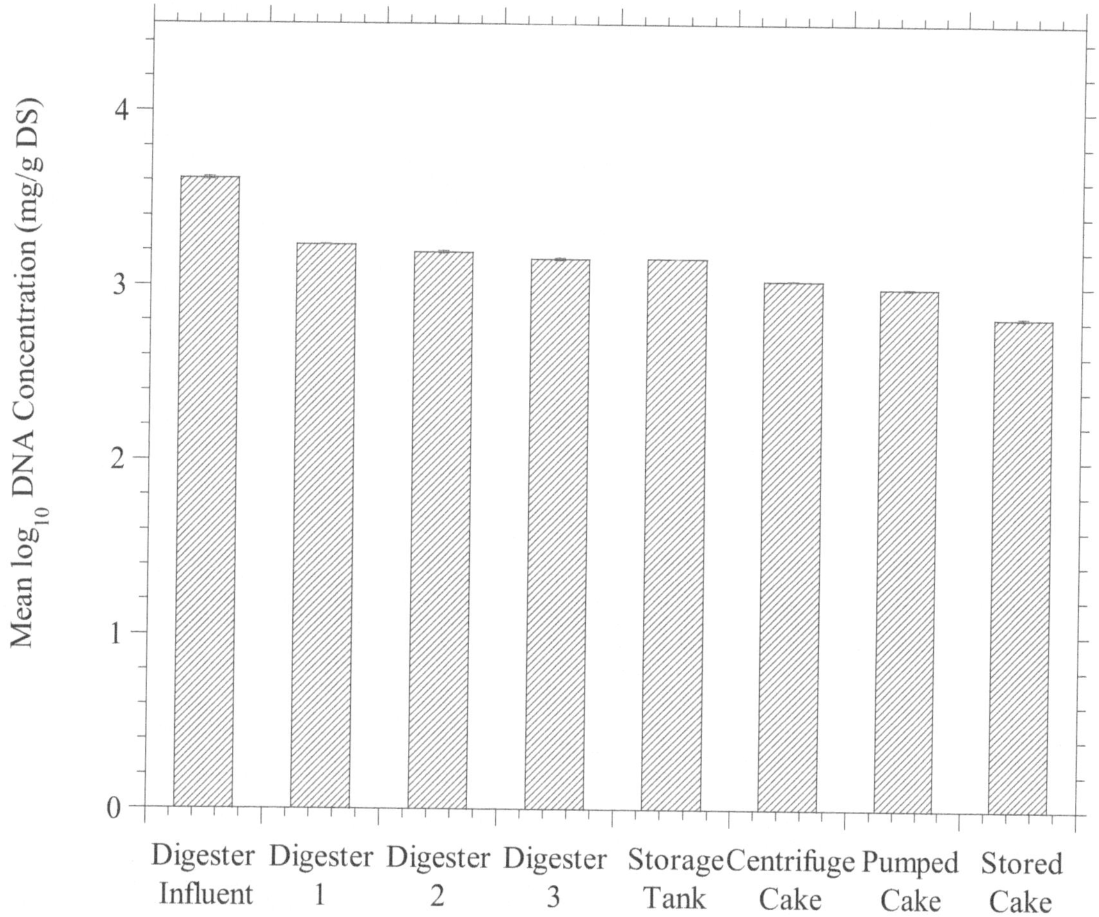

Figure 3-6. DNA Concentration Measured in the Thermophilic Plant 2 Digestion and Dewatering Process. (Error Bars Represent One Standard Deviation).

3.1.3 Thermophilic Plant 2 – Dewatering from Different Stages

The results from the research have shown that multi-stage or reactors-in-series may be important in achieving desired indicator destruction. To study this further, a large-scale field trial was performed in which the operation of Thermophilic Plant 2 was modified such that dewatering could be performed with only one, two or three stages plus the storage tank. Unfortunately, due to the site limitations, dewatering had to be performed directly from the storage tank, but the Stage 2 and/or Stage 3 reactor could be taken off-line. As a result, at least two reactors in series were also being used, but the storage tank depth was lowered such that the retention time was approximately 0.5 days for the experiments, and then on the last sampling event the storage tank was operated at approximately 1 d. A schematic of the reactor configuration for each sampling event is shown in Figure 3-7. After each change in operation, which entailed removing one reactor from operation, the system was operated for two to three days to achieve steady state conditions, and then the samples were collected. The entire sampling event was performed over about a two week period. Samples were collected for enumeration of FC and *E. coli* by both SCM and qPCR as in previous work. In addition, cake samples were collected for regrowth studies.

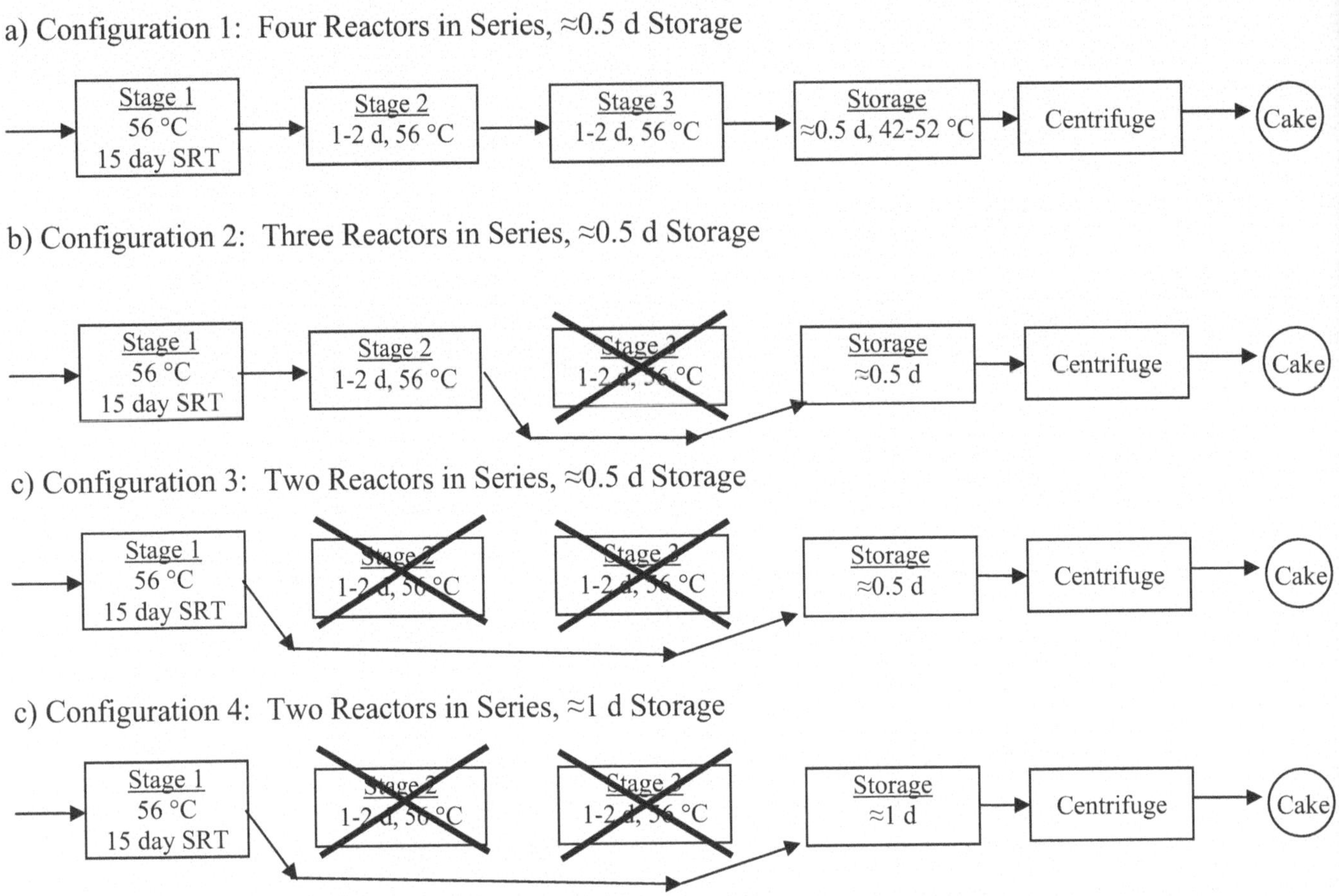

Figure 3-7. Schematic of the Different Reactor Configurations Studied in the Thermophilic Plant 2 Field Trial.

The first day of sampling entailed collection from all reactors in series, and the results are shown in Figure 3-8. This sampling event was essentially a repeat of the Thermophilic Plant 2 sampling that was discussed in Section 3.1.2, with a difference in time of approximately ten months. Interestingly, the influent density of FC was about one order of magnitude greater than previous sampling, and in this event, detectable densities of FC were enumerated in the digested samples analyzed by SCMs. In addition, unlike the previous event discussed in Section 3.1.2, the density of *E. coli* was greater than the detection limit for the RT-PCR method. The *E. coli* densities were in the range of 10^4 to 10^5 cells/g DS. However, similar to the previous sampling event, no SI occurred, in other words, no significant increase in the FC density was measured in the cake samples immediately after dewatering. Unfortunately, regrowth was not studied in any of the sampling events for this trial due to shipping problems, mainly the samples were held up in U.S. Customs for at least one extra day and in some cases more. This resulted in holding times much greater than that recommended by the U.S. EPA (1999), and the samples were therefore not analyzed. This eliminated both the regrowth data sets and the *E. coli* enumeration by the SCMs. Therefore, the data set is limited to analysis of SI of FC after centrifugation. The research team was very disappointed in this outcome, since such a great effort was put into this event. Fortunately, some data was still collected and it provides some additional insights.

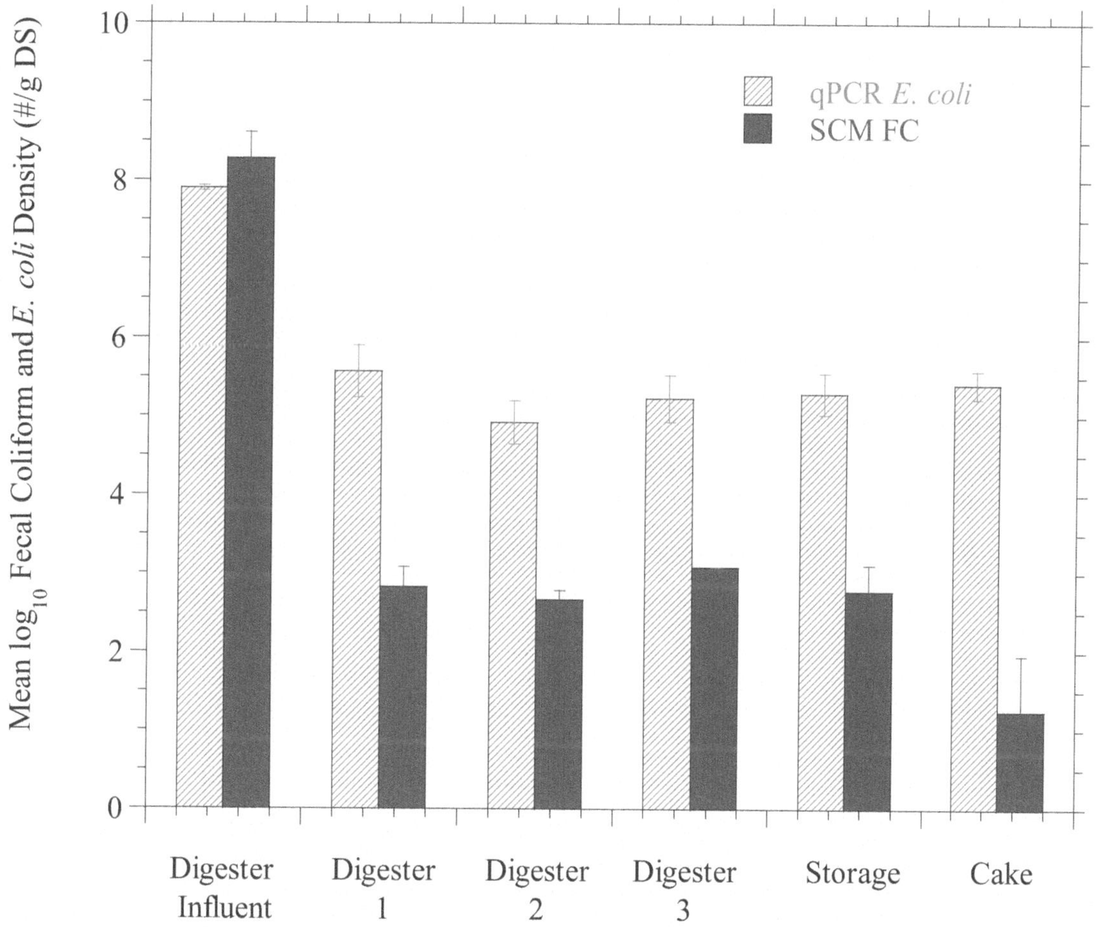

Figure 3-8. FC and *E. coli* Density Through the Thermophilic 2 Digestion and Dewatering Process as Measured by Culturing Methods and RT-PCR. (Error Bars Represent One Standard Deviation).

The RT-PCR results for each of the different configurations were very similar in terms of *E. coli* density measured from the reactor effluents. The FC density measured in the cake samples for each configuration is shown in Figure 3-9. Interestingly, as the number of reactors in series decreased, the FC density increased. However, the increases were modest, and the SI potential (RT-PCR minus the SCM in the digester effluent) of about 3 to 4 log difference, was not met for any of the reactor configurations. This suggests a large proportion of the *E. coli* DNA measured by RT-PCR may not have represented *E. coli* that were able to be reactivated.

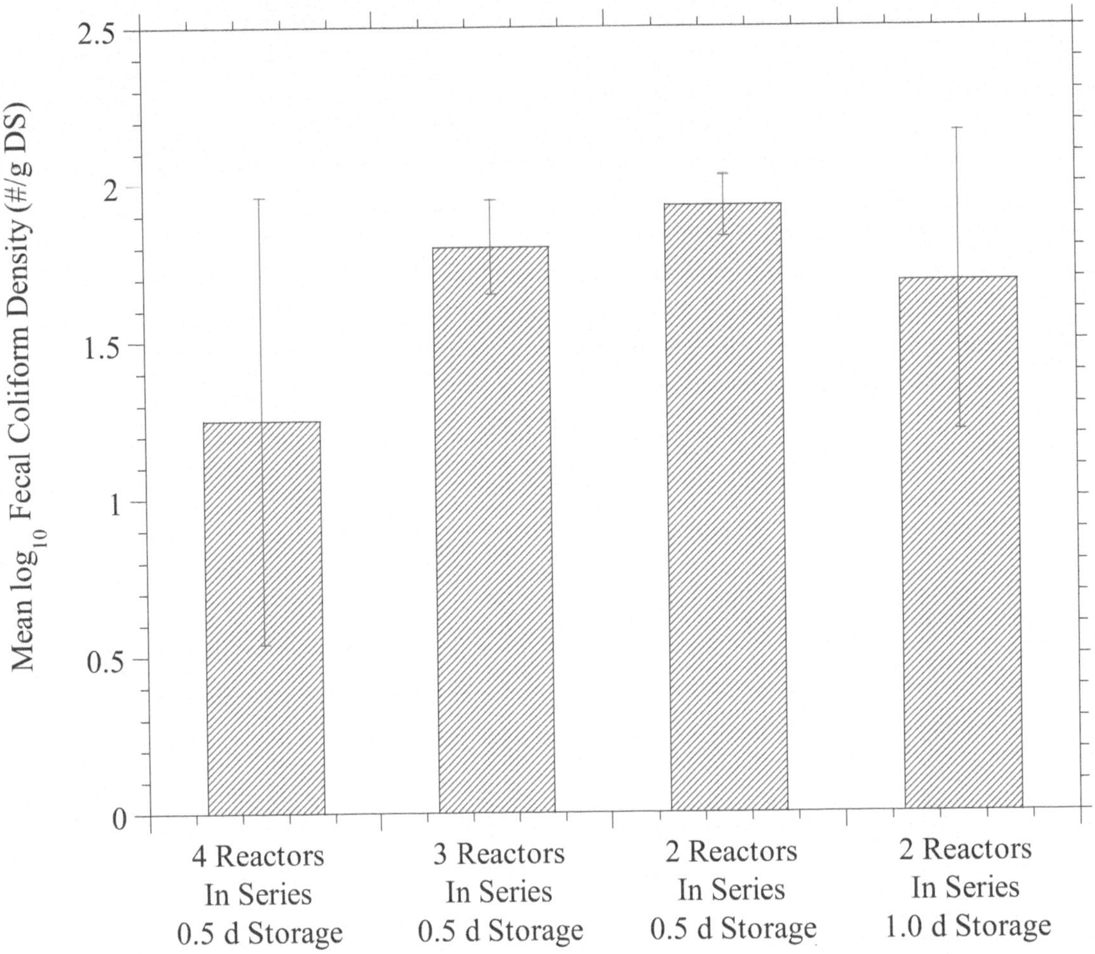

Figure 3-9. Mean FC Density Measured in Cake Samples for the Different Reactor Configurations.

The greater amounts of *E. coli* measured by the RT-PCR method during this sampling event compared to several previous events which were below the detection limit are interesting. For this sampling event, the influent densities were about one order of magnitude greater than previous events. This may have resulted in an increase in the effluents due to greater densities entering the digestion process. As a result, the quantities of *E. coli* were greater than the detection limit of the qPCR method, and were therefore able to be enumerated. The previous events may have had *E. coli* densities measured by qPCR that were just below the detection limit and therefore not enumerated.

3.1.4 Pre-Pasteurization with Mesophilic Digestion.

This plant utilizes pre-pasteurization to achieve the time-temperature required for Class A biosolids. The pertinent operational parameters are provided in Table 3-3. The process uses the U.S. EPA batch time-temperature equation to achieve Class A requirements, followed by conventional mesophilic digestion. Dewatering is performed with high solids centrifugation, with typical cake solids in the range of 27-29%. After centrifugation, the cake is transported via a scroll conveyor to cake storage hoppers. Sampling of each process was performed, and the FC and *E. coli* densities measured by SCM as well as the *E. coli* measured by qPCR are shown in Figure 3-10. Prior to pre-pasteurization, the FC and *E. coli* densities were in a typical range around 10^7/g DS, and the qPCR and SCM methods agreed. After pre-pasteurization and prior to digestion, the SCM densities decreased to below the detection limit; however, the qPCR for *E. coli* remained consistent with the density prior to the treatment. After digestion, the SCM densities remained below the detection limits, and the qPCR density of *E. coli* decreased by about two orders of magnitude, to about 10^5/g DS. After dewatering, the SCM measured densities increased by about four orders of magnitude, and the SCM densities agreed well with the qPCR measured densities. The sample taken in the storage bin showed about one order of magnitude greater density, around 10^6/g DS for both the SCM and qPCR measurements. The sudden increase immediately after dewatering was in the range of 4-5 orders of magnitude (depending on the indicator). It is likely that this level of increase cannot occur in that short of a time period as a result of just regrowth because the doubling time of *E. coli* is on the order of 20 minutes for ideal conditions.

Table 3-3. Summary of Digester Operational Parameter for Pre-Pasteurization Process.

Parameter	Pre-Pasteurization	Mesophilic Digester
Primary/Secondary Ratio	60/40	-
Digester SRT	Minimum of 45 minutes-typ. 2 hrs	29 d
Digester Temperature	66°C	35°C
Mixing	Yes	canon mixer
Digester pH	-	7.2
Total Solids	4.3%	1.5%
Volatile Solids	78%	58%
Volatile Solids Destruction	-	-
Cake Solids	-	28%

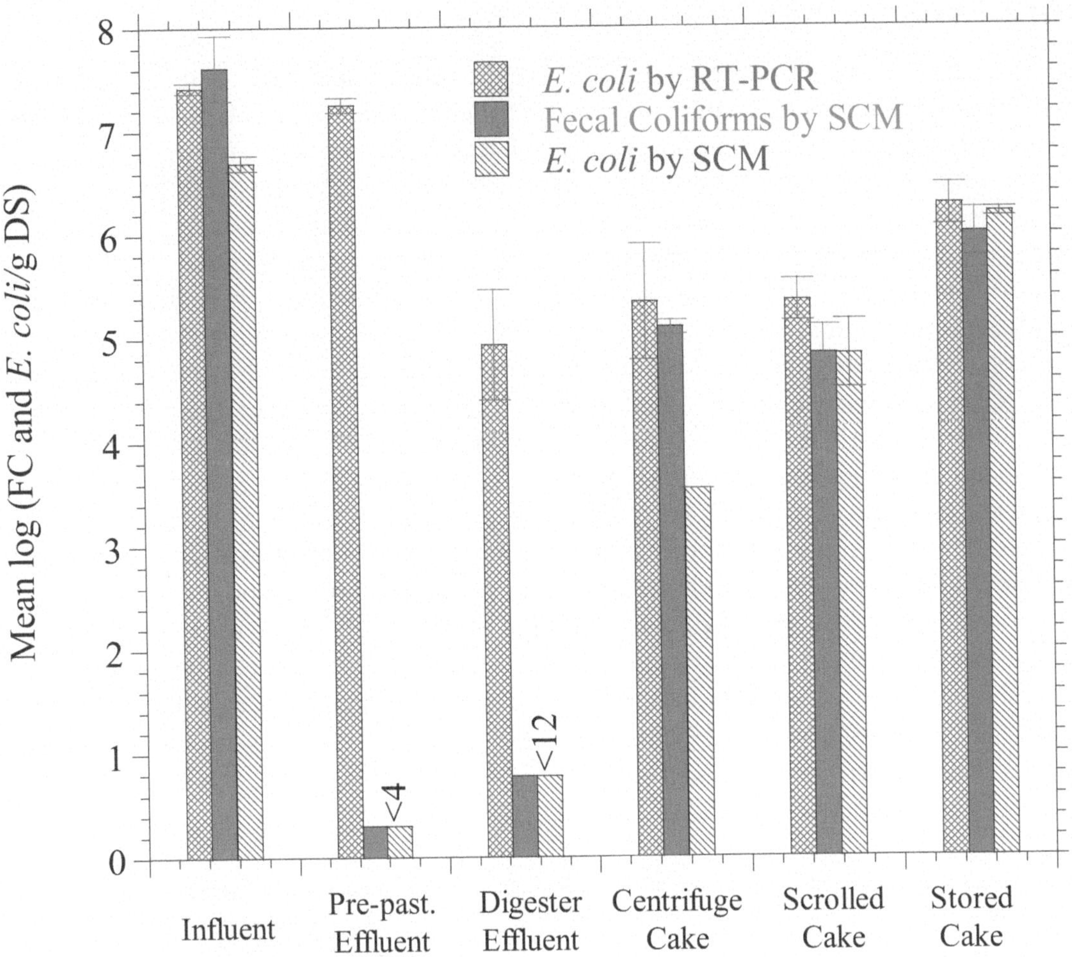

Figure 3-10. FC and *E. coli* Densities Measured in the Pre-Pasteurization, Digestion and Dewatering Process. (Error Bars Represent One Standard Deviation).

3.1.5 Mesophilic Plant 1

This plant utilizes two, single stage, mesophilic anaerobic digesters, followed by storage in holding tanks, and dewatering is performed on high solids centrifuges. The relevant operational parameters are provided in Table 3-4. Dewatering is performed with a Humboldt/Bird high solids centrifuge which produced a mean cake solids of 24.3% during the month of the sampling event. The mean cationic polymer dosage was 9.5 kg/tonne.

WERF

Table 3-4. Summary of Digester Operational Parameter for Mesophilic Plant 1.

Parameter	Digester 1	Digester 2	Storage
Primary/Secondary Ratio	ND	ND	-
Digester SRT	31 d	32 d	≈ 1 d
Digester Temperature	38.0	38.0	ND
Mixing	Gas mix = draft tubes	Gas mix = draft tubes	Pump mix
Digester pH	7.2	7.2	ND
Digester Alkalinity	4222 mg/L	4383 mg/L	ND
Digester Volatile Acids	403 mg/L	409 mg/L	ND
Total Solids	1.79%	1.83%	ND
Volatile Solids	73%	74%	ND
Volatile Solids Destruction	56%	56%	ND

[1]ND=no data available

The results showed that the bacteria do not appear to enter a non-culturable state during digestion to later be reactivated during dewatering. This is shown by the lack of increase in FC by the MPN method before and after dewatering, Figure 3-11. In addition, the *E. coli* measurements showed that the qPCR results are very similar to the SCM results which are shown in Figure 3-12. This is additional evidence that the bacteria did not enter the non-culturable state in significant numbers. In other words, the SI potential, taken as the difference between the qPCR density and the SCM density, was low.

It is interesting to note that significant regrowth occurs after dewatering and pumping the cake with piston pumps followed by several days of storage in cake silos at the plant; this suggests that the release of substrate may occur during centrifugation which enables growth during storage. This data is discussed in more detail in the Section 3.2 on "Regrowth"

The total DNA extracted from each sample is shown in Figure 3-13. The digestion removed about 28% of the DNA compared to the influent sample. The large drop in DNA concentration in the storage bin (after centrifugation) may be due to death of anaerobic organisms caused by exposure to oxygen.

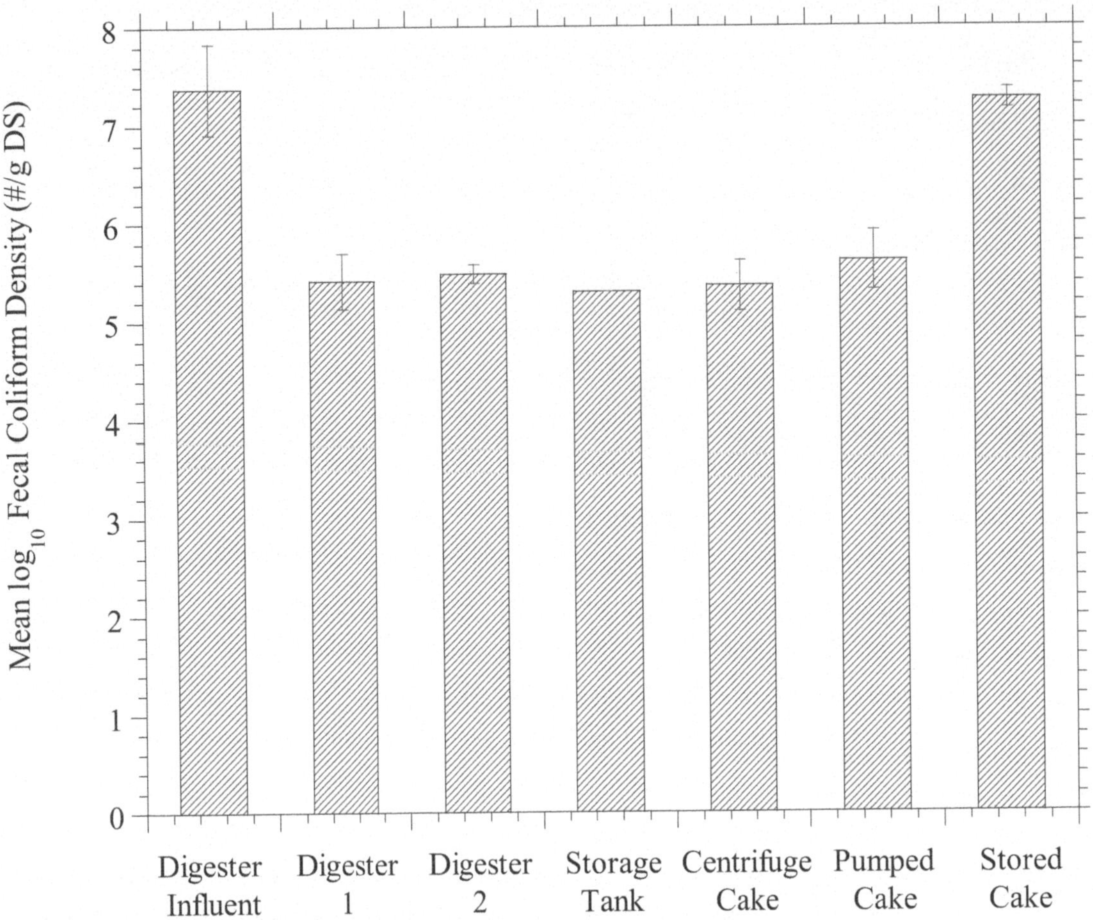

Figure 3-11. FC Densities Through the Mesophilic 1 Digestion and Dewatering Processes. (Error Bars Represent One Standard Deviation).

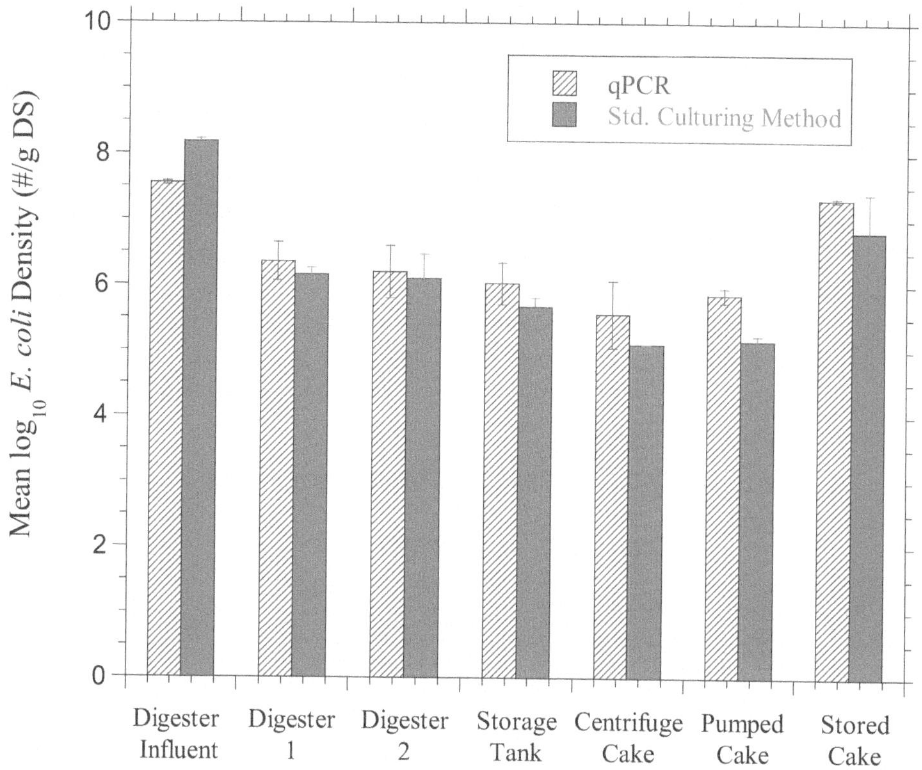

Figure 3-12. ***E. coli*** **Density Through the Mesophilic 1 Digestion and Dewatering Process as Measured by Culturing Methods and RT-PCR. (Error Bars Represent One Standard Deviation).**

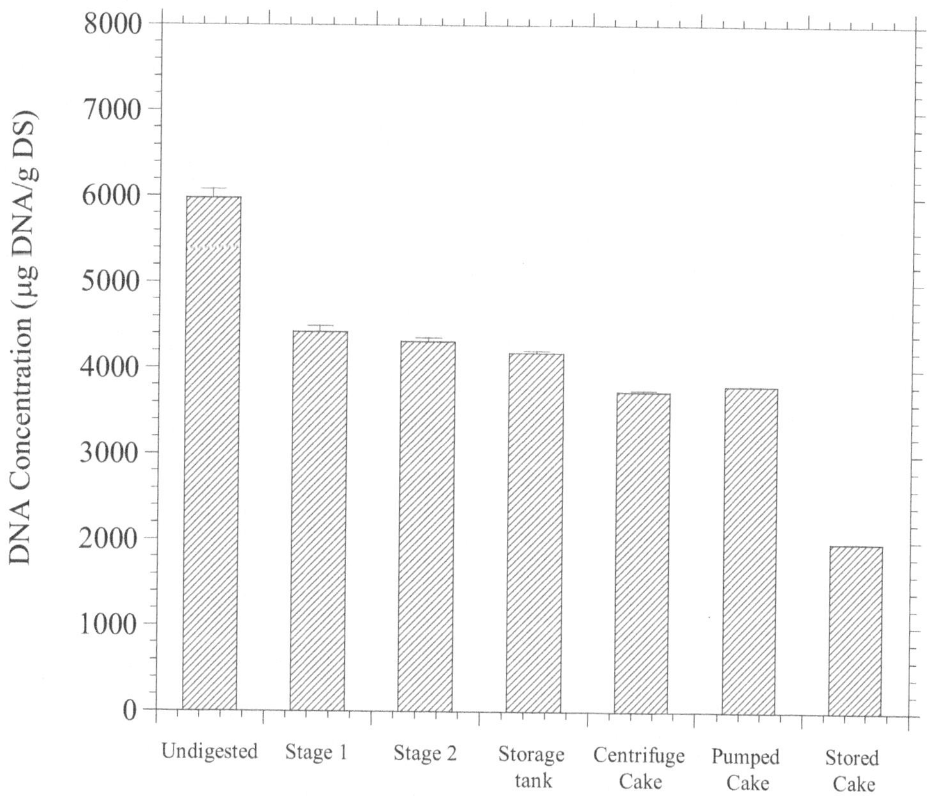

Figure 3-13. DNA Concentration Measured in the Mesophilic Plant 1 Digestion and Dewatering Process. (Error Bars Represent One Standard Deviation).

3.1.6 Mesophilic Plant 2

Mesophilic Plant 2 utilizes two stage (in series) mesophilic digestion, followed by thickening and dewatering on a belt filter press. A summary of the operating parameter for Mesophilic Plant 2 is provided in Table 3-5. The FC and *E. coli* results from both SCM and RT-PCR are shown in Figure 3-14. The cell densities for the indicator organisms decline during the two stage digestion as well as during thickening and storage of the solids prior to dewatering by the belt filter press. Similar to the thermophilic digestion processes, the reactors in series approach appears to provide better reduction in the indicator organism. No SI was measured after belt filter press dewatering, and from the RT-PCR and SCM densities measured during storage, it appears the SI potential was relatively low.

Table 3-5. Summary of Digester Operational Parameter for Mesophilic Plant 2.

Parameter	Digester 1	Digester 2	Storage
Primary/Secondary Ratio	40/60	-	-
Digester SRT	15 d	10-15 d	7-14 d
Digester Temperature	36°C	36°C	ND
Mixing	Recirculation Pump	Recirculation Pump	ND
Digester pH	7-7.4	7-7.4	ND
Digester Alkalinity	2500-3500 mg/L	2500-3500 mg/L	ND
Digester Volatile Acids	ND	190	ND
Total Solids		1.5%	4%
Volatile Solids	65-66%	60-65%	ND
Volatile Solids Destruction	60%		

[1]ND=no data available

The RT-PCR method produced an unusually high and unexpected *E. coli* density after BFP dewatering, and it is not clear if this was due to an error in measurement, or potential contamination of the sample or some other reason. Regrowth did not occur during storage of the cake as discussed in Section 3.2.

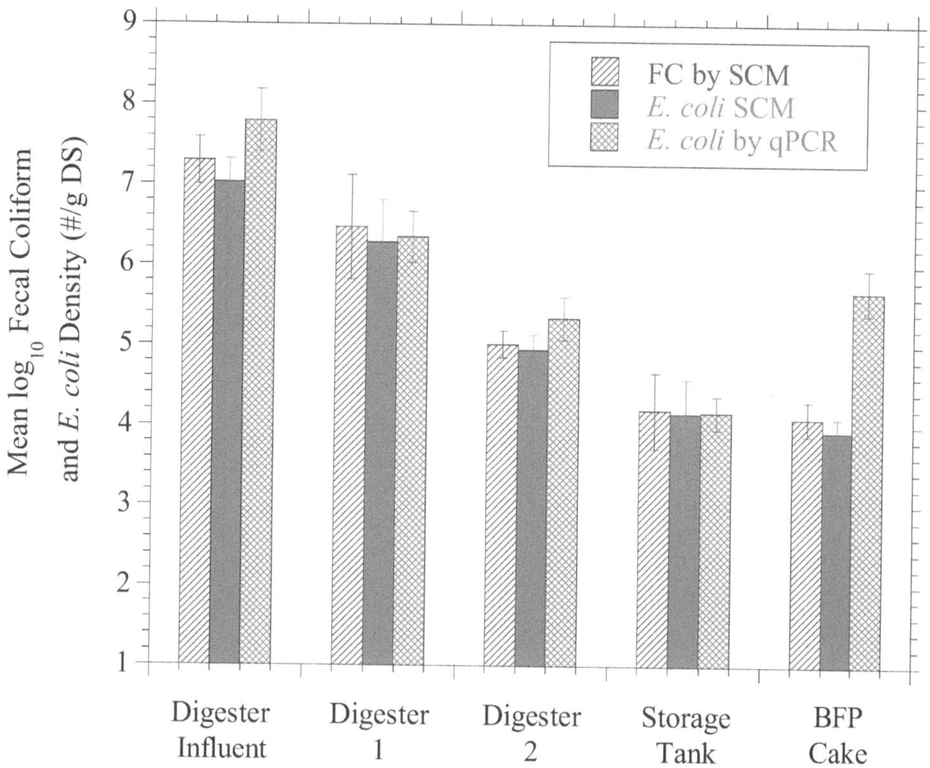

Figure 3-14. FC and *E. coli* Densities Measured in the Mesophilic Plant 2 Digestion and Dewatering Process. (Error Bars Represent One Standard Deviation).

3.2 Regrowth During Cake Storage

3.2.1 Centrifuge Dewatering.

A set of regrowth and die-off experiments were performed on cake samples collected from Thermophilic Plants 1 and 2, and Mesophilic Plant 1 which all had centrifuge dewatering. The cake samples after centrifugation were collected and stored in the laboratory at mesophilic temperatures. During regular intervals, sub-samples were taken and analyzed for FC and *E. coli* using SCMs. The results from these storage experiments are shown in Figures 3-15 and 3-16. For Thermophilic Plant 1 and Mesophilic Plant 1, the FCs and *E. coli* increased rapidly during storage and reached a peak concentration after one day of storage. Regrowth was defined to occur when the increase in density was statistically significant ($p<0.05$) when comparing the density at day 0 with the highest density measured. For both Thermophilic Plant 1 and Mesophilic Plant 1, the increase was statistically significant, and it was not for Thermophilic Plant 2.

The peak density of both FC and *E. coli* approached 10^8 cfu/g DS for both samples with regrowth. Interestingly, Thermophilic Plant 2 did not have any SI or growth during storage. After reaching a peak concentration, the FC and *E. coli* densities decreased during the next two months of storage for the two samples that showed regrowth. The rate of decrease was greater for the mesophilic plant compared to the thermophilic plant. It is not clear why the mesophilic cake had faster die-off, compared to the thermophilic cake, although it may be related to the greater numbers of bacteria that compete for available substrate. The cake samples from Mesophilic

Plant 1 had about eight times (3800 µg/g DS) the DNA concentration measured in the cake from Thermophilic Plant 1 (550 µg/g DS) which suggests greater numbers of microbes were present in the mesophilic cake.

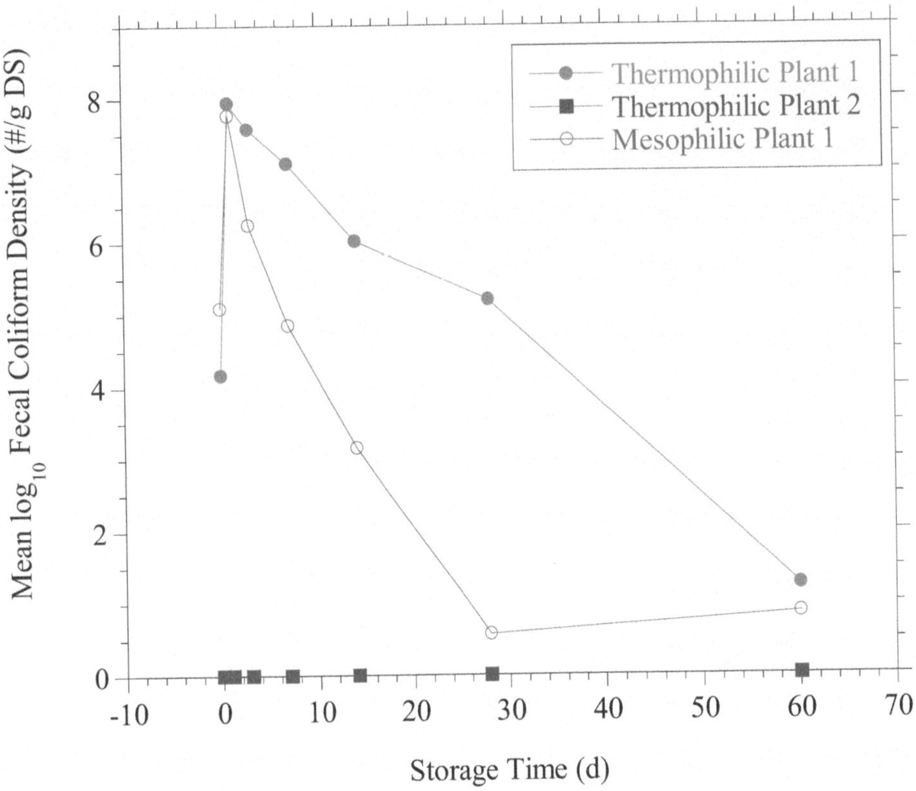

Figure 3-15. Mean FC Densities during Storage of Cake in the Laboratory.

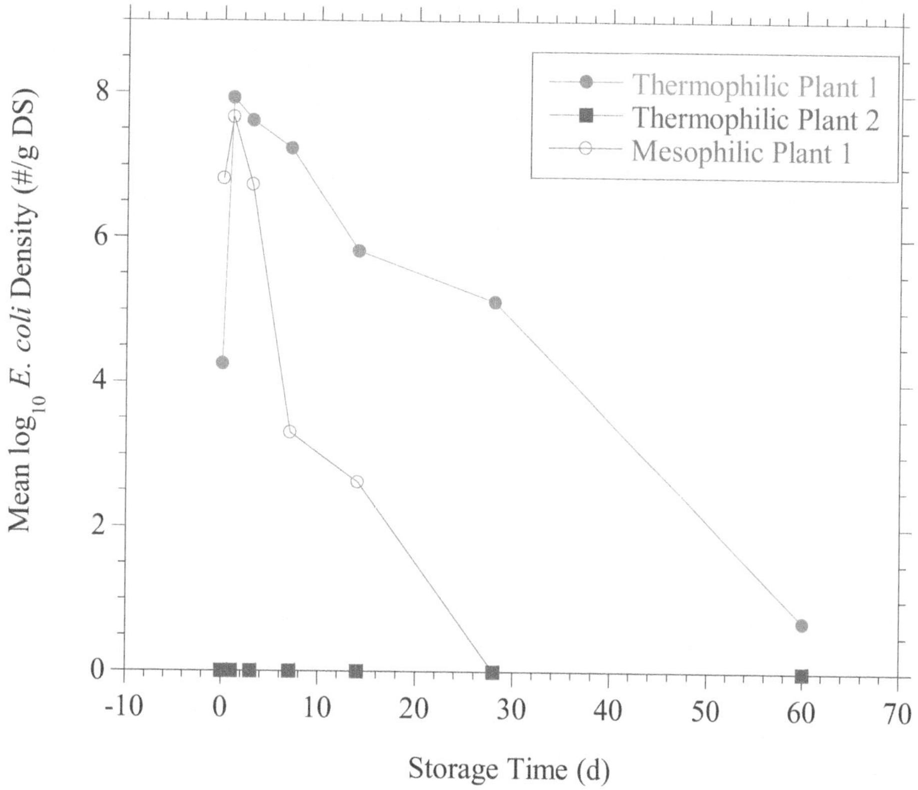

Figure 3-16. Mean *E. coli* Densities during Storage of Cake in the Laboratory.

It is interesting to note that regrowth did not occur in the cake of Thermophilic Plant 2. This could be a result of the lack of FC and *E. coli* in the cake to begin with since this plant had the lowest densities measured by all methods during the sampling event. In addition, the digestion process may have reduced the available substrate that could support regrowth. An experiment was performed to determine if regrowth would occur in Thermophilic Plant 2 cake when it was spiked with *E. coli*. This spiking would provide the seed of *E. coli*, and if conditions were suitable in the cake, regrowth would be expected. During storage over 48 hours, no significant increase in the *E. coli* density was measured in the cake after spiking (data not shown), suggesting that the conditions in the cake were not suitable to support growth. It may also be possible that the presence of inhibitors from digestion may reduce regrowth as well.

Spiking of the cake from Thermophilic Plant 1 resulted in a slight increase in the rate of regrowth; however, the peak density of *E. coli* was approximately equivalent after 48 hours for both the spiked and unspiked sample (data not shown).

3.2.1.1 Effect of Shear After Centrifuge Dewatering

A second set of samples was taken from Mesophilic Plant 1 to examine the impact of shear due to cake pumps on the growth during storage. Figure 3-17 shows the mean FC densities measured in cake samples taken immediately after centrifugation, and also after transport through a cake pump downstream of the centrifuge. The additional shear associated with the cake pump did not have any significant impact on growth. For cake samples, the peak concentration as well as the die-off curves were very similar.

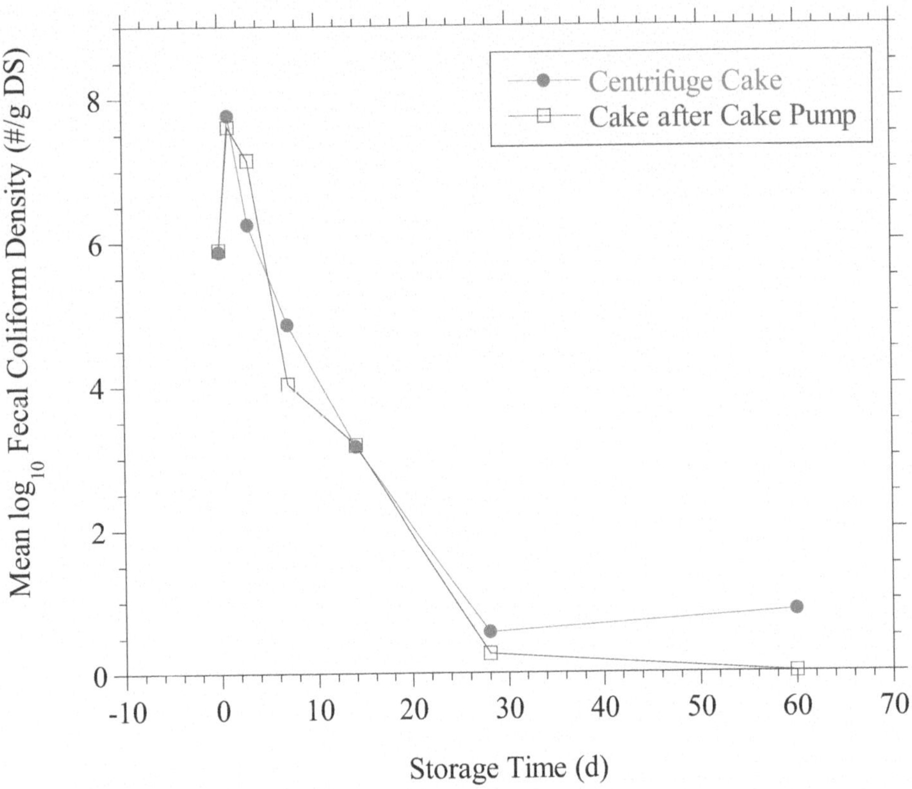

Figure 3-17. Mean FC Concentration of Mesophilic Plant 1 Cake Sample Directly After Centrifugation and After Pumping Through a Cake Pump.

In a similar study, the impact of cake storage in on-site storage silos was examined for Thermophilic Plant 2. The FC densities measured in the cake collected after cake pumps as well as cake that had been stored for several days in a silo are shown in Figure 3-18. The results showed that after the on-site storage, FCs were present in the cake, but at fairly low concentrations, with all samples meeting Class A biosolids regulations for FC. The results suggest that contact with the on-site storage facility may have seeded some FCs into the cake, although a rapid and sustained growth did not occur after seeding. This may be due to lack of available substrate, presence of inhibitors, and/or competition from other microbes.

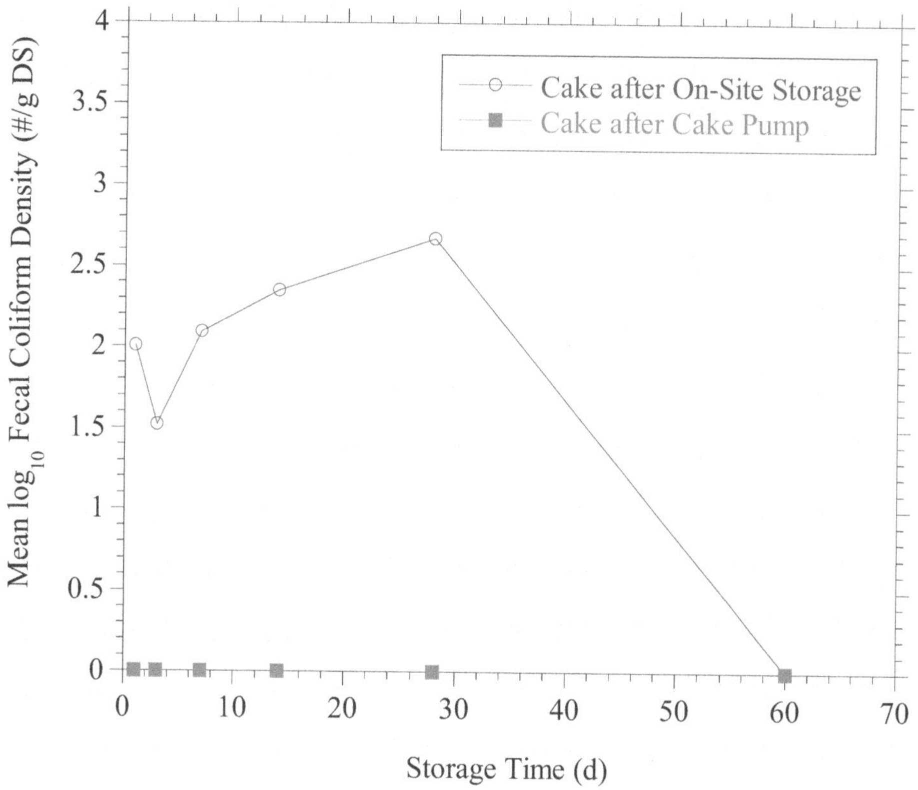

Figure 3-18. FC Densities in Laboratory Stored Cake Samples Taken from Thermophilic Plant 2 Before and After On-Site Storage.

3.2.1.2 Comparison of High and Medium Solids Centrifuges

Another set of samples were collected at a mesophilic plant that utilized several different centrifuge types of dewatering. The purpose of this experiment was to determine if SI and regrowth occurred with high and medium solids centrifuges. This plant, Mesophilic Plant 3, has both Humboldt high solids centrifuges and modified, low solids Bird centrifuges. The high solids centrifuge achieves cake solids in the range of 30-33% solids, while the medium solids centrifuges produce cakes in the range of 25-30%. The terms for medium and high solids are relative and are defined for this plant with these solids ranges. The regrowth curves for both cake samples are shown in Figure 3-19. Cake samples from both centrifuges had similar FC densities at the initial time, and their regrowth and die-off pattern were very similar

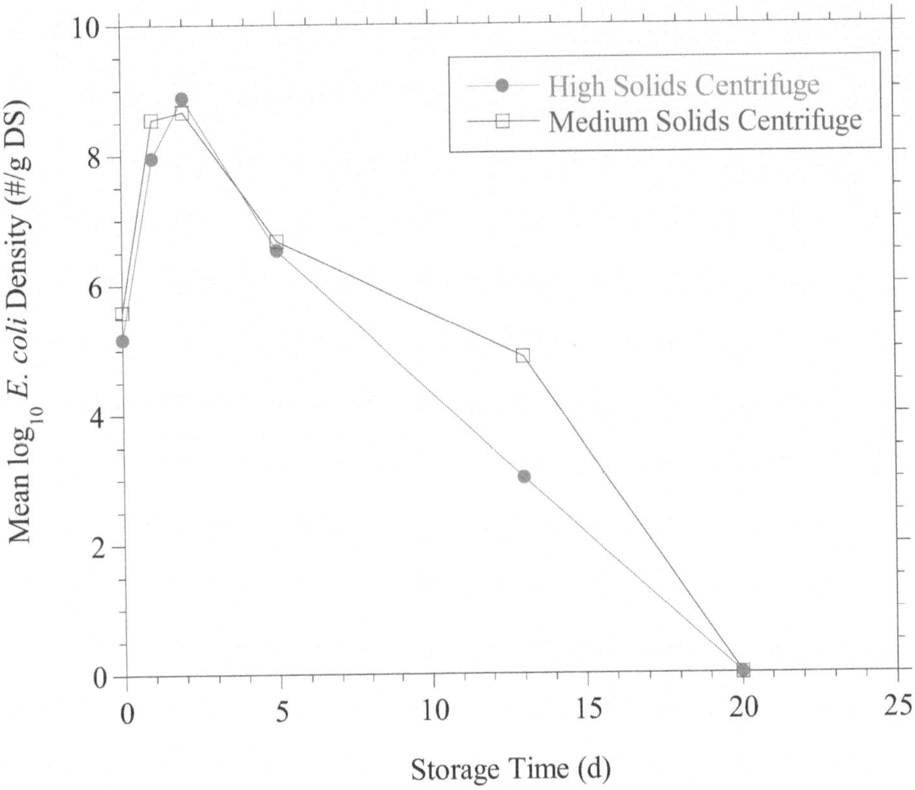

Figure 3-19. FC Densities in Laboratory Stored Cake Samples Taken from Mesophilic Plant 3 from Two Different Centrifuge Types.

3.2.1.2 Effect of Shear Within a Centrifuge

Samples were collected for a field trial in coordination with another WERF project on odors in biosolids. These samples were collected from a pilot centrifuge dewatering mesophilic anaerobically digested biosolids. During dewatering, the centrifuge was stopped, and the scroll was removed from the bowl. Cake samples were then collected along the length of the scroll and sent to HML for FC and *E. coli* enumeration. The locations of the samples were taken in sequence along the length of the scroll as follows:

- Location 1 – Flight 3-4 along the horizontal portion of scroll
- Location 2 – Flight 7-8 along the horizontal portion of scroll
- Location 3 – Flight 8-9, about halfway up the inclined cone
- Location 4 – Discharge chute of the centrifuge (centrifuge effluent).

The effect of the sampling location on the FC and *E. coli* densities measured during cake storage is shown in Figure 3-20 and 3-21, respectively. No qPCR data was collected for this trial. The results showed that as the samples first entered the centrifuge and traveled along the horizontal portion of the bowl, some SI/regrowth occurred, but very little difference was measured from the two locations along this horizontal portion of the scroll (Locations 1 and 2). The cakes solids did increase between Locations 1 and 2 as shown in the graphs. The cake samples taken in the conical section (Location 3), had a much higher cake solids of 32% compared to Location 2. Also, the amount of SI/regrowth increased compared to the previous two samples. Similarly, the cake taken at the discharge of the centrifuge (Location 4) had even a higher level of SI/regrowth, and the cake solids was lower by about 4% points. These results give support to the hypothesis that shear is important to SI and regrowth. Also, the data indicates that much of the shear occurs in the entrance and exit zones of the centrifuge.

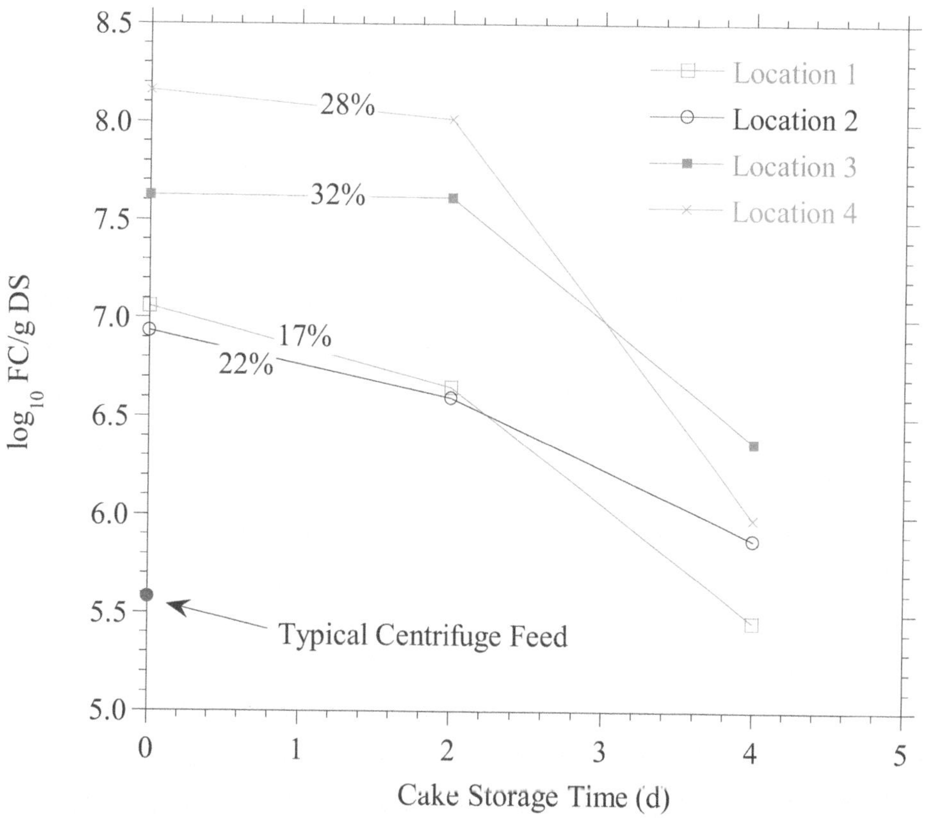

Figure 3-20. FC Densities Measured During Storage of Cake Samples Taken Along the Scroll of a Field Centrifuge, the Numerical Values Shown on the Lines are the Respective Cake Solids for Each Sample.

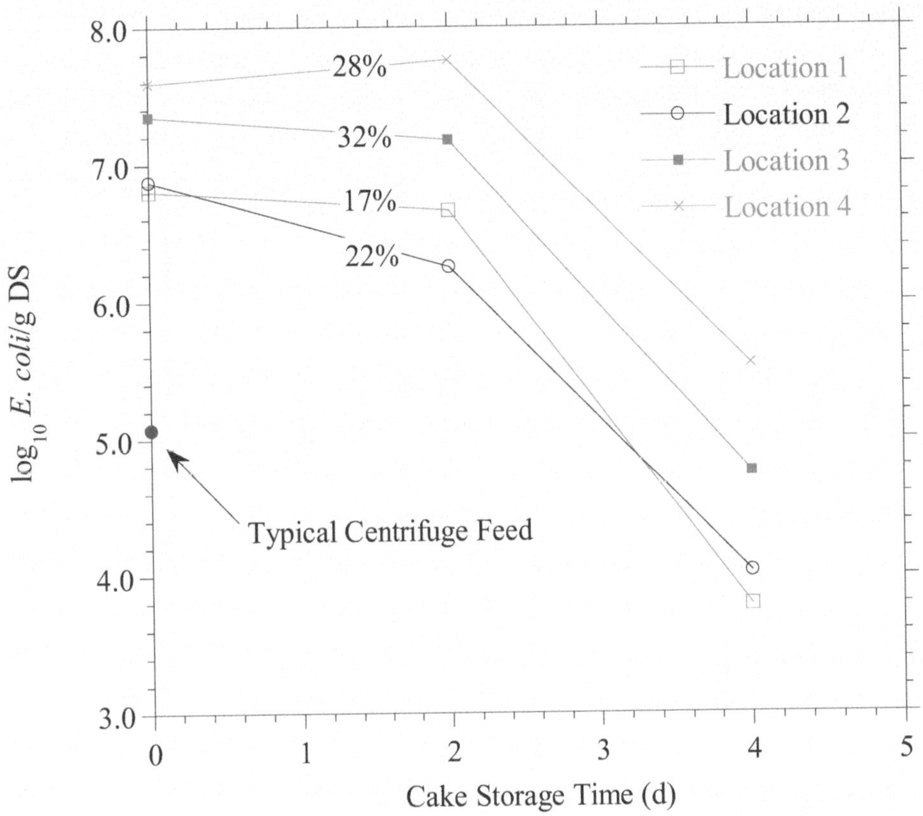

Figure 3-21. ***E. coli*** **Densities Measured During Storage of Cake Samples Taken Along the Scroll of a Field Centrifuge, the Numerical Values Shown on the Lines are the Respective Cake Solids for Each Sample.**

3.2.2 Belt Filter Press Dewatering

Three cake samples were collected from belt filter press dewatering processes, with all the samples being digested with mesophilic anaerobic digestion. A summary of the operating parameters for the three plants is provided in Table 3-6. The FC during cake storage for the three samples is shown in Figure 3-22. No substantial regrowth was measured in these belt filter press cakes, although slight increases were measured in Mesophilic Plant 2 and 5's cake samples. However, this increase was much less than amount measured during storage of centrifuge cake samples when regrowth occurs. Interestingly, the die-off curves varied considerably for each plant. No significant ($p<0.05$) increase in the FC and *E. coli* densities were measured during storage as shown in Figures 3-22 and 3-23. The statistical analysis was performed by comparing the FC and/or *E. coli* density on day 0 with the highest density measured during cake storage. If the increase is statistically different at $p<0.05$, then regrowth was said to occur.

Table 3-6. Summary of Digester Operational Parameter for Mesophilic Plants 2, 4, and 5.

Parameter	Mesophilic Plant 2	Mesophilic Plant 4	Mesophilic Plant 5
Digester Configuration	Two in Series	Two in Series	Two in Series
Total Digester SRT	25-30 d	70 d	24 d
Digester Temperature	36°C	37.5°C	40°C
Mixing	Recirculation	Recirculation	Gas
Digester pH	7.2	7	7
Digester Alkalinity	3000 mg/L	2500 mg/L	ND
Digester Volatile Acids	190 mg/L	<150 mg/L	ND
Total Solids	1.5%	1.3%	2.2%
Volatile Solids	62.5%	67%	60%
Volatile Solids Destruction	60%	57%	48%

[1]ND=no data available

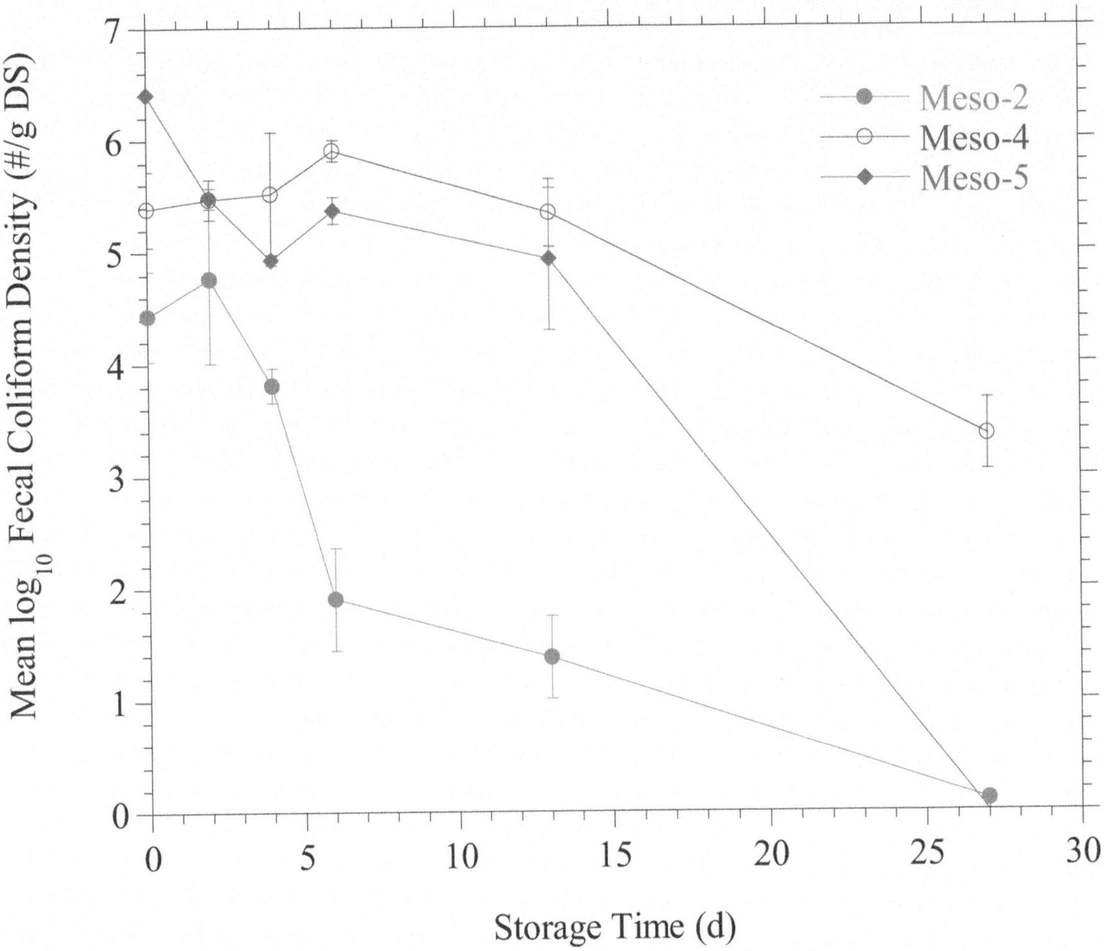

Figure 3-22. FC Densities in Laboratory Stored Cake Samples Taken from Three Different Mesophilic Plants Using Belt Filter Press Dewatering. (Note: The upper error bar represents one standard deviation of the mean value).

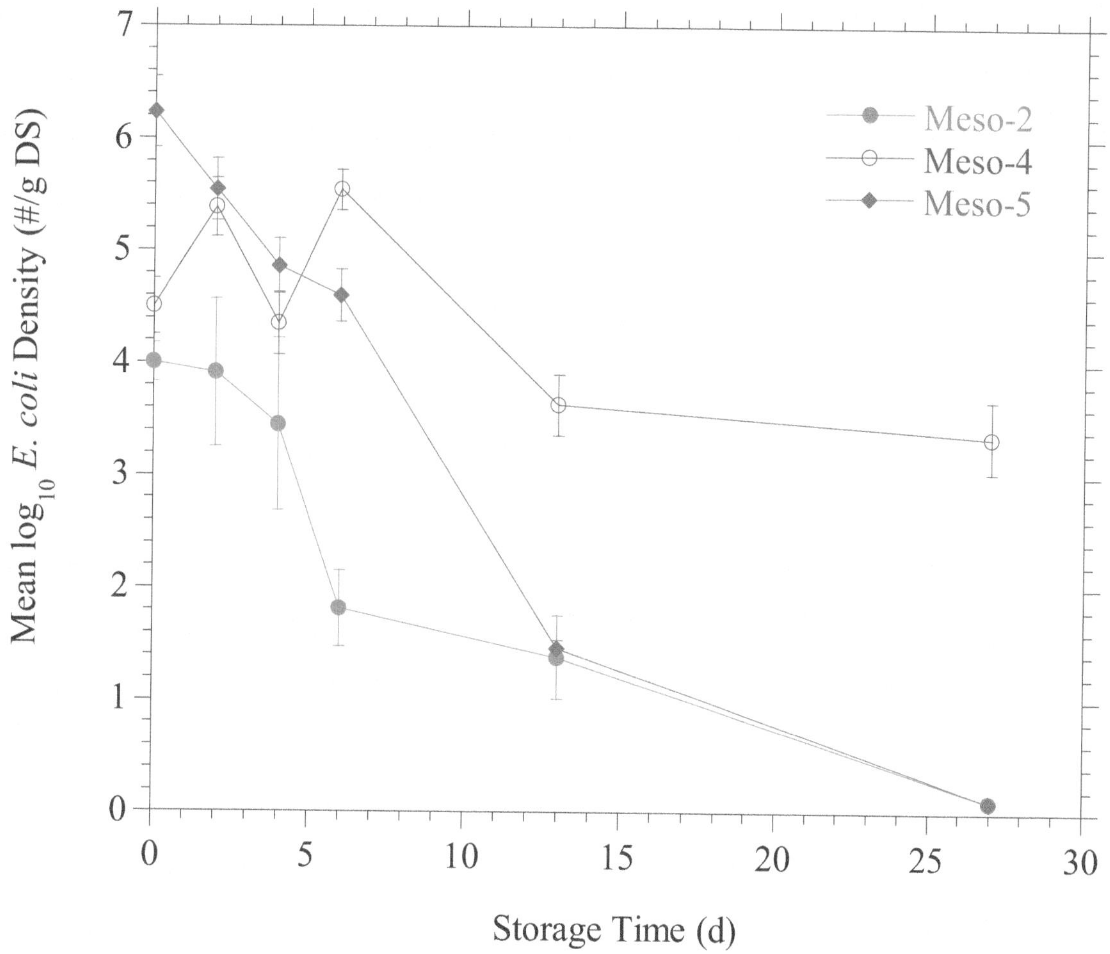

Figure 3-23. ***E. coli*** **Densities in Laboratory Stored Cake Samples Taken from Three Different Mesophilic Plants Using Belt Filter Press Dewatering.**

3.2.3 Rotary Press Dewatering.

A sample set was also collected from a full-scale trial with a Rotary Press. The Rotary Press is a relatively new technology for dewatering. Cake samples from a Rotary Press were collected and analyzed for FC and *E. coli* during storage. The results are shown in Figure 3-24. The Rotary Press dewatering showed an increase in the FC densities, however, not the *E. coli* densities.

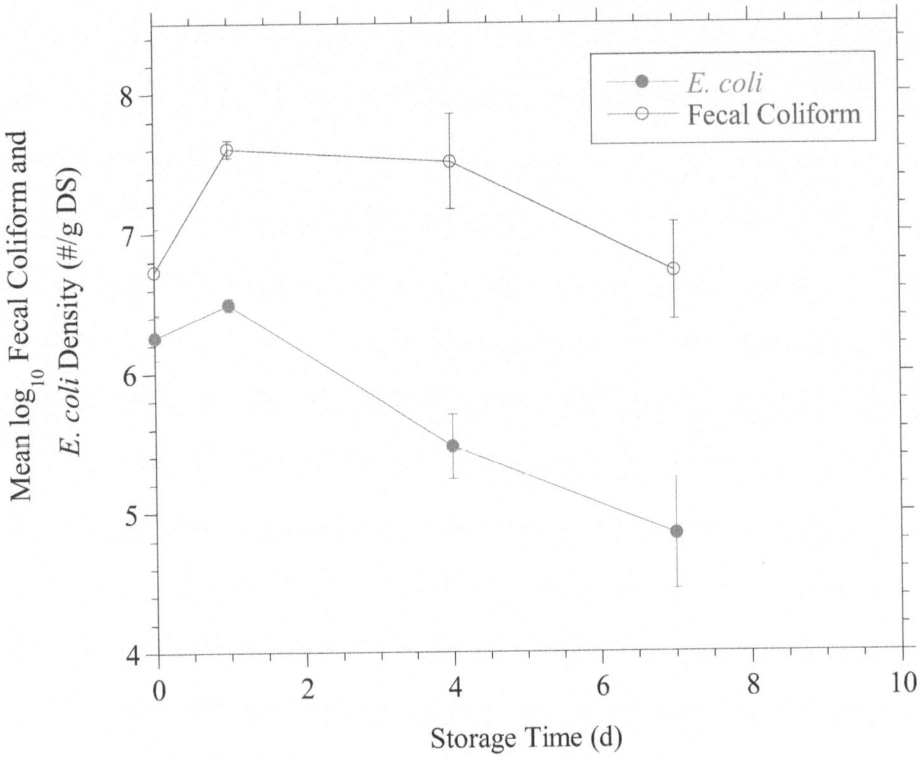

Figure 3-24. Growth of FC and *E. coli* in Cakes Produced Using a Rotary Press.

3.2.4 Undewatered Digested Solids.

As a control test, an experiment was performed to determine if regrowth occurs during storage of mesophilically digested solids that were not dewatered. Digester effluent samples were collected on two different weeks (one week apart) from Mesophilic Plant 3, prior to dewatering. The density of FC and *E. coli* were measured during storage of the sample for seven days at 35°C. The FC density measured in the liquid digested solids is shown in Figure 3-25, and the *E. coli* densities for the same experiment are shown in Figure 3-26. The results showed that FC and *E. coli* decreased steadily during storage of the liquid sample at 35°C, and confirm that the regrowth in samples is not an artifact of sampling procedures and storage in the laboratory.

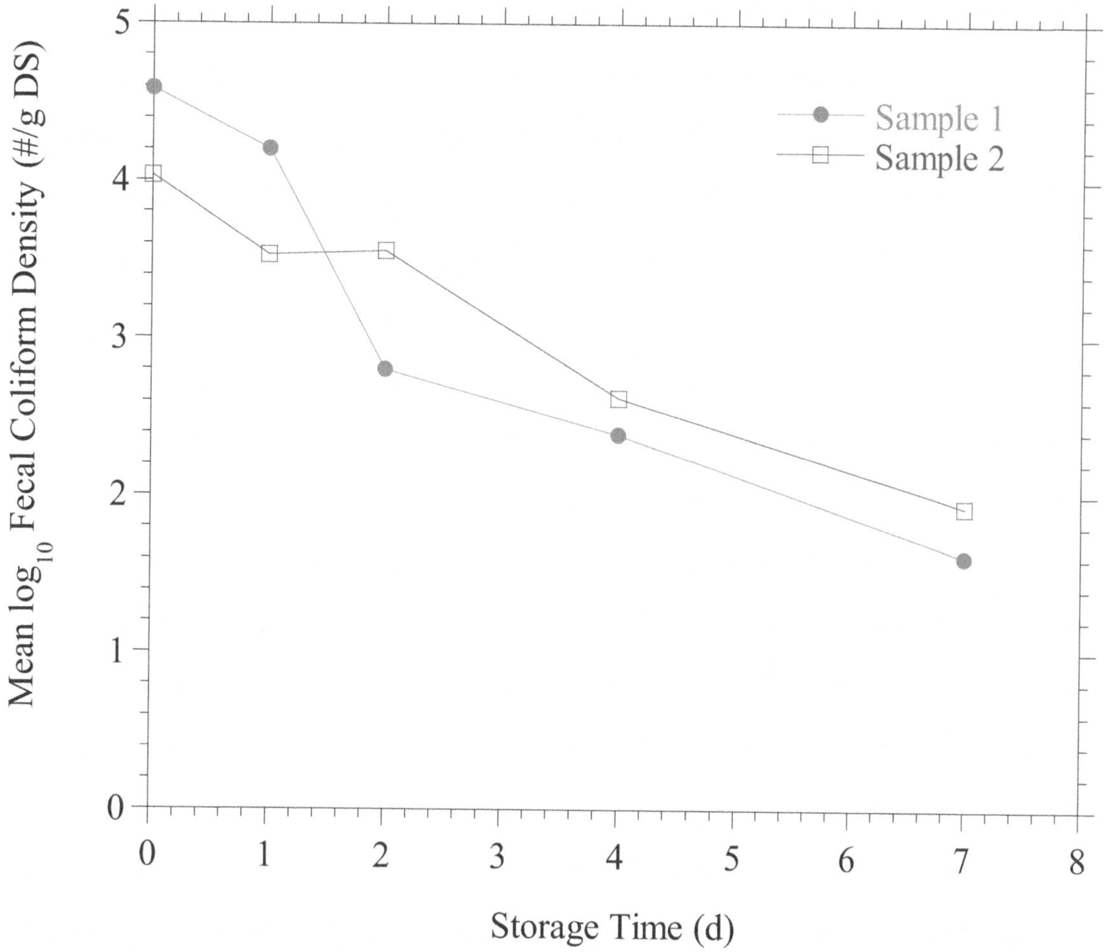

Figure 3-25. FC Density in Mesophilic Plant 3 Liquid Sludge during Batch Storage.

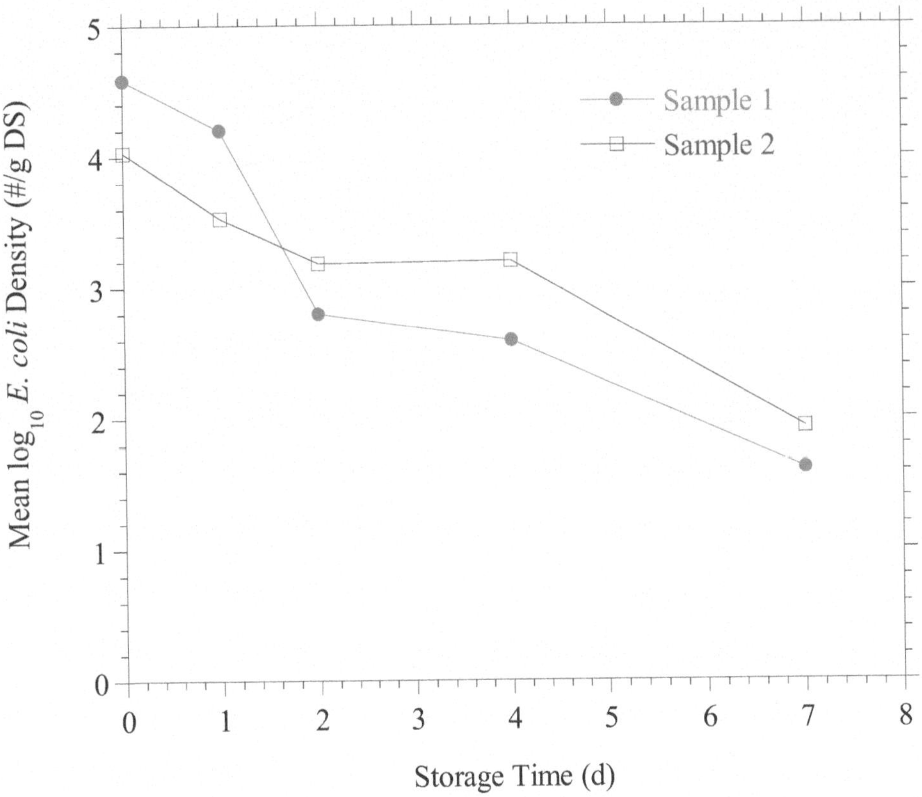

Figure 3-26. ***E. coli*** **Density in Mesophilic Plant 3 Liquid Sludge during Batch Storage.**

3.2.5 Effect of Temperature on Regrowth

A set of experiments were performed to determine the impact of temperature on regrowth rates during 48 hours of storage. Samples were collected from Thermophilic Plant 1, and were stored at 37°C and room temperature (about 22°C). The effect of storage temperature on *E. coli* density measured in the cake during the first 48 hours is shown in Figure 3-27. No difference in the rate of regrowth was found in this experiment, suggesting that temperature was not limiting the growth of the *E. coli*. Similar results were found for the heterotrophic plate counts as shown in Figure 3-28. This may suggest that the growth was substrate limited rather than temperature limited.

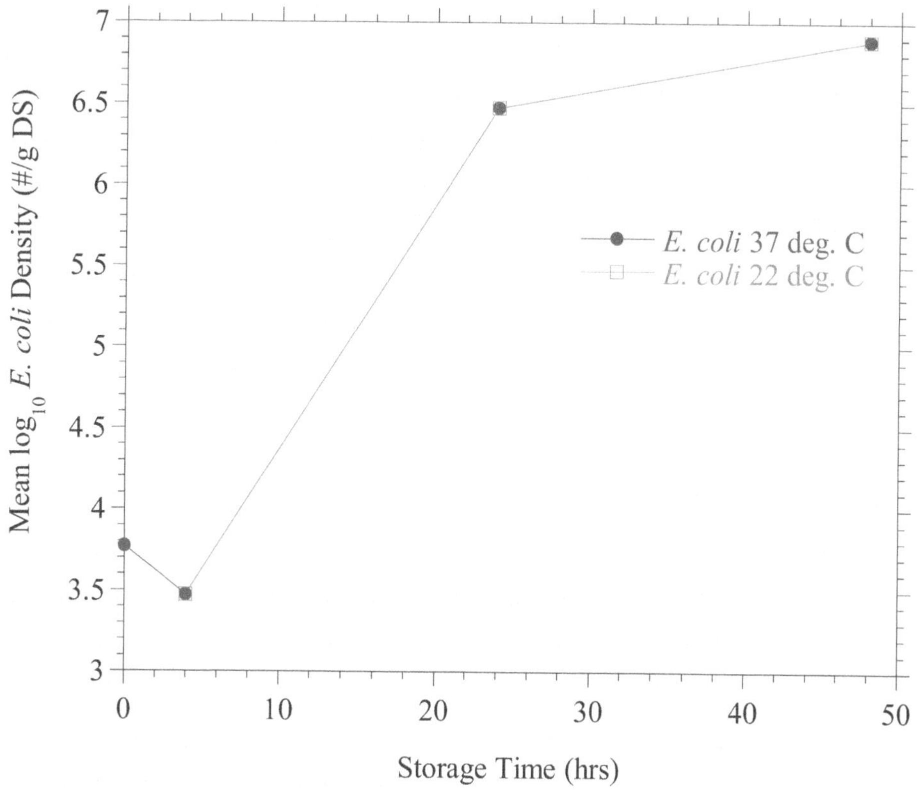

Figure 3-27. Effect of Storage Temperature on *E. coli* Regrowth in Thermophilic Plant 1 Cake Samples After Centrifugation.

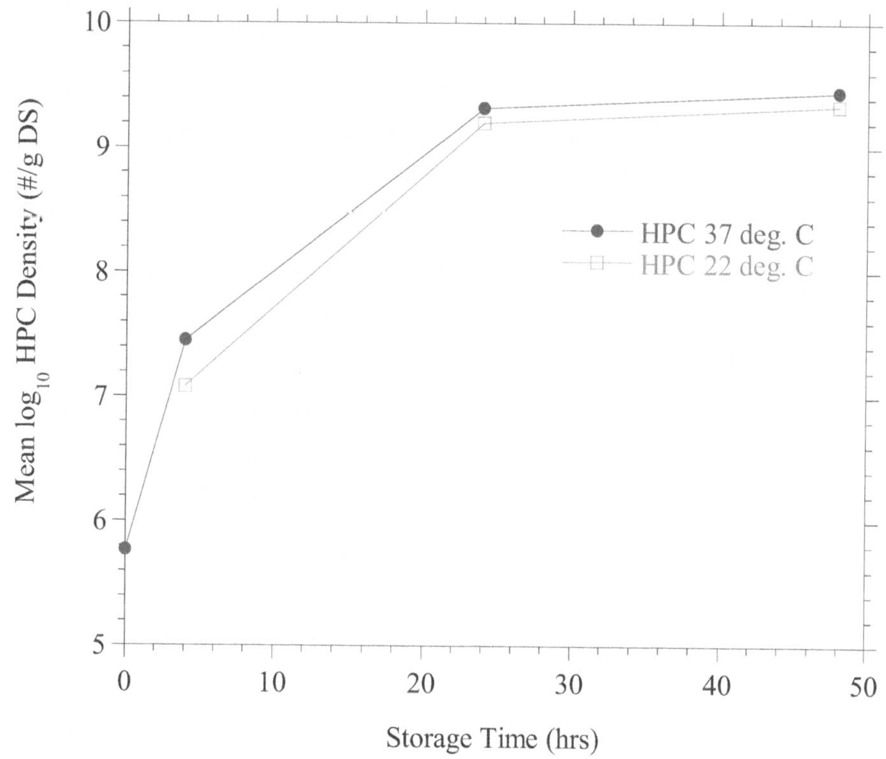

Figure 3-28. Effect of Storage Temperature on HPC Regrowth in Thermophilic Plant 1 Cake Samples After Centrifugation.

3.3 Discussion and Evaluation of Field Data

The following section attempts to evaluate both the data from this study, as well as other studies in the literature. In addition, several utilities have provided data for the project team to evaluate. The literature and these field experiences help to broaden the data set and allow for a comprehensive evaluation of the SI and regrowth phenomenon. The first section (3.3.1) examines the "field experiences", which is data provided by utilities for the purpose of evaluating and including in the analyses. In some cases, the utilities and their agents have also graciously provided a write-up for their data for us to include. Due to confidentiality agreements, the names of all the people who have helped are not revealed, however the WERF Project Team thanks the many people who have helped with this project.

The next section (3.3.2) examines the literature that is available related to SI and regrowth. All this available data are then synthesized to evaluate possible trends and correlations between operational parameters and SI/regrowth in Section 3.3.3.

3.3.1 Field Experiences

Several utilities have provided data that they have collected related to FC after digestion and/or dewatering. The data helps to broaden the data set for SI and regrowth information. It should be stressed that the project team did not help to collect or analyze the samples; therefore, the quality control aspects used in this research project may or may not have been equivalent in these field experiences. Despite this limitation, evaluation of this data is still instructive.

3.3.1.1 Autothermal Thermophlic Aerobic Digestion

The treatment plant in question is a 9000 m^3/day municipal wastewater treatment facility that uses a trickling filter/solids contact (TF/SC) process following primary treatment to create a secondary treatment effluent. The primary solids are thickened in the primary clarifiers whereas the secondary solids are thickened using an internally-fed rotary drum thickener. The volume of thickened primary solids is typically in the range of 24 m^3/day at about 5.6% dry solids whereas the volume of thickened secondary solids is typically in the range of about 6 m^3/day at about 6.6%. The resulting blended sludges are about 5.7% to 6.3% dry solids, but typically in the 6% range. The blended sludges are fed to an autothermal thermophilic aerobic digester (ATAD) system for volatile solids and pathogen destruction.

The ATAD system consists of a series of six sealed, but connected, rectangular concrete tanks, each with approximately 135 m^3 of potential liquid volume. These tanks are mixed and aerated using Turborator™ vertical aspirating mixers, two per tank. Initially, three ATAD tanks were used, in series, with feed of the blended thickened sludges going into Tank 4 and then being displaced into Tank 5 (and Tank 5 into Tank 6). Using temperature probes and a variable frequency drives to control the aeration rates, these tanks were kept at about 40°, 50°, and 60°C, respectively. Based on the average flow rate of about 932 m^3/month, the average hydraulic retention time (HRT) per tanks would be about 4.5 days per tank or 13.5 days overall for three tanks in series, which equates to about 675° days at an average of 50°C. While the local requirements for Class A biosolids expect that at least 5 days at 50°C will achieve Class A pathogen kills (i.e. less than 1000 fecal coliforms per dry gram of biosolids), this was not always achieved. It was thought that since the tanks are hydraulically connected, there could be some short-circuiting that may be occurring at times. Also, since the aerators are the mixers, when the mixers are turned down to decrease aeration and temperatures, there could be some settling of heavy materials, including grit, which would tend to build up in the corners of the rectangular tanks, decreasing the effective HRT. On this basis, two changes were made to the system: A fourth ATAD tank, Tank 3, was brought into service and the new ATAD was equipped with the latest generation of Turborators that can vary the aeration rate while maintaining their mixing energy.

After a trial period when the temperatures in the four ATAD tanks were 30°, 40°, 50°, and 60°C, respectively, the temperature regime was set up as 40°, 50°, 60°, and 60°C, respectively. This final regime eliminated any concerns about pathogen kills. A round of fecal coliform testing on November 20, 2006 provided the results shown in Table 3-7 below:

Table 3-7. Results of Fecal Coliform Testing for the ATADs, November 20, 2006.

Sample Location	Temperature (°C)	Results as reported MPN/100 wet g	Results as MPN/ wet g	Percent dry solids	Results as MPN/ g dry solids
Raw mixed sludge	18	1,100,000	11,000	6%	183,333
ATAD 3	40	>16,000,000	>160,000	5%	>3,200,000
ATAD 4	50	460,000	4,600	3.1%	148,390
ATAD 5	60	<200	<2	4.6%	<43.5
ATAD 6	60	<200	<2	3.4%	<58.8

As can be seen from Table 3-1, the FC kills are progressive through the ATAD system and, seemingly, very effective in achieving Class A requirements of less than 1000 fecal coliforms per dry gram.

Once the sludges have completed their treatment in the ATAD system, they are dewatered and hauled away for land application as Class A biosolids. The initial method of dewatering was using a 1.5 m-wide belt filter press (BFP). While this BFP was reasonably effective as a dewatering device, the exposure and splashing of ATAD-treated biosolids resulted in the release of many difficult-to-treat odor compounds including methyl mercaptans, dimethyl sulphide and dimethyl disulphide. Since there was no redundancy for the single BFP, when the opportunity and budget came available for upgrading the dewatering system, there was no question that the new dewatering device would be a centrifuge.

The existing dewatering building was modified and expanded to permit the installation of new Alfa Laval G2-60 centrifuge as well as a new dry polymer system. The dry solids content of the cake increased to the 30% and higher dry solids range in very short order and, as a result, everyone was happy. The plant staff had a much easier and cleaner dewatering system to use, the local residents did not have to worry as much about the odors from ATAD biosolids processing and the biosolids hauling contractor had much smaller volumes to worry about. It seemed like a win-win-win.

That is, until some of the fecal coliform tests on the newly centrifuge-dewatered cake came back non-Class A. For example, one result was 920,000 MPN/100 wet g (about 31,000

MPN/ dry g) off an influent sludge from ATAD 6 with less than the lab's 20 MPN/100 wet gram detection limit (about 6 MPN/dry g). The first thing that was done was to resample and retest to confirm. The results were much the same. The treated sludge coming out of Tank 6, going into the centrifuge, met the Class A pathogen indicator requirements but the dewatered cake coming out of the centrifuge had pathogen indicators well above the Class A requirements. The next thing was to test the existing belt filter press (BFP) against the centrifuge. Since the BFP was never decommissioned and was intended as the back-up to the centrifuge, this was relatively easy. The results were very clear, as shown in Table 3-8.

Table 3-8. Comparison of Fecal Coliforms after Dewatering with BFP and Centrifuge.

Sample Location	Fecal Coliforms (MPN/100 wet g)	Fecal Coliform (MPN/ wet g)	Percent dry solids	Results as MPN/ g dry solids
ATAD 5	230	2.3	4.6%	50
ATAD 6	<20	<0.2	3.4%	<5.9
Belt Filter Sample 1	<2000	<20	20%	<100
Belt Filter Sample 2	<2000	<20	20%	<100
Centrifuge Sample 1	79000	790	30%	2,633
Centrifuge Sample 2	540,000	5400	30%	18,000

As can be seen from Table 3-8, the feed to the dewatering devices met Class A requirements with respect to viable pathogens, at least based on the fecal coliform indicator results (i.e. <1000 MPN/dry g). It was also clear that the ATAD solids dewatered by the BFP still met the Class A pathogen indicator requirement. However, the centrifuged ATAD solids did not. While the two samples have an order of magnitude difference, even the better of the two samples still fails to meet the Class A pathogen indicator requirement.

Following the confirmation that the centrifuge was causing some sort of SI of pathogen indicators, the next thing that was done was to test to see if the speed of the centrifuge had any effect (i.e. whether slowing the centrifuge down would be beneficial in decreasing the SI). In combination, it was decided to test whether there would be pathogen indicator die-off if samples were taken on one day and then stored at 20°C (in a BOD incubator) over a series of days). The results of this testing, in MPN /dry gram are presented in Table 3-9.

Table 3-9. Investigation of Fecal Coliform Die-off After Centrifuging an ATAD Sludge.

Date of sample submission (date of initial sampling was Dec. 5th)	3250 RPM Sample MPN/dry g	2800 RPM Sample MPN/dry g
Dec.5/06	30,667	11,667
Dec. 11/06	53,333	53,333
Dec. 15/06	53,333,333	11,666,667
Dec. 20/06	5,333,333	566,667
Jan. 03/07	1,166,667	3,066,667

As can be seen from Table 3-9, there appears to be some minor benefit in running the centrifuge at a slower speed. However, the difference is not enough to allow the ATAD biosolids to remain Class A, as they were before dewatering. The die-off is also relatively slow, with high densities being measured even after about one month of storage.

The next round of testing will involve slowing the centrifuge down further and testing for dry solids and viable fecal coliforms in the dewatered cake to determine whether there is a "sweet spot" where the dewatering is still acceptable, e.g. 22 to 25% dry solids, and the indicator organisms still meet the Class A requirements.

3.3.1.2 Anaerobic Mesophilic Digestion with Centrifuge Dewatering

This plant utilizes mesophilic anaerobic digestion, followed by centrifuge dewatering. The plant did a trial in which samples of the centrifuge feed and centrifuge cake were collected. The cake samples were stored and sampled at regular intervals. At the time of the sampling, the digestion SRT was 22 days, the digester temperature was 36°C, and the VS reduction during digestion was 45%. After digestion, the solids content was 2.2%, and this was centrifuge dewatered to a cake solids of 24%. The FC densities measured in the feed and cake are shown in Figure 3-29. The results did not show an immediate SI, however, significant regrowth was measured after one day of storage.

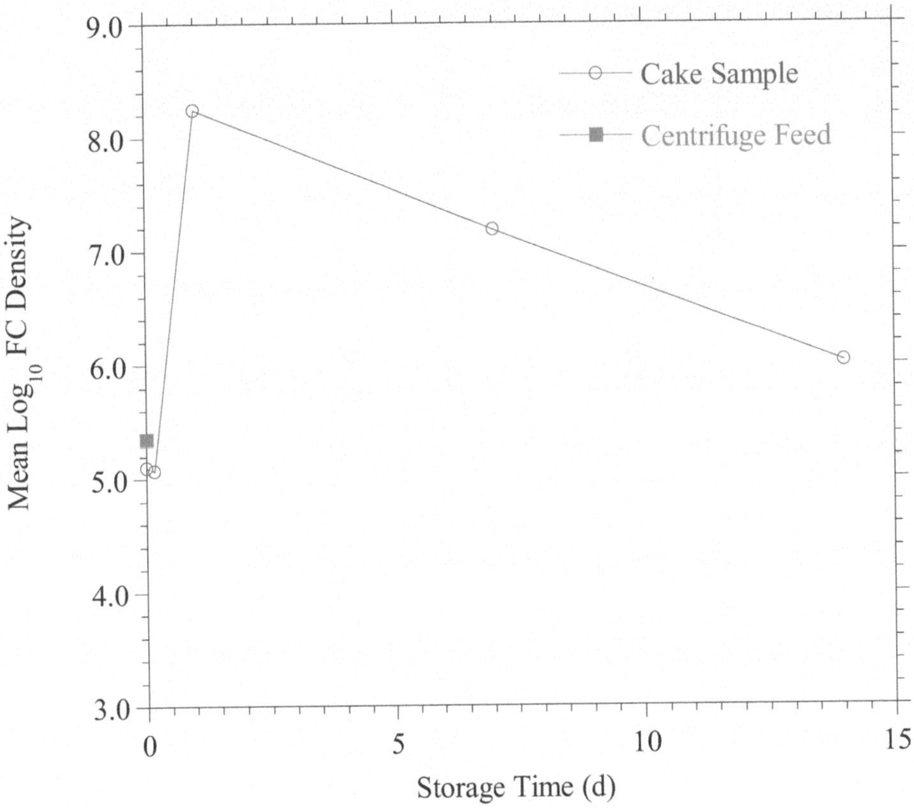

Figure 3-29. FC Density Measured by SCM After Mesophilic Digestion and Centrifuge Dewatering as well as During Cake Storage.

3.3.2 Evaluation of Field Results and Literature

Based on the data available from this study, as well as the literature (reviewed in Chapter 1.0), the total number of data points is 44. It should be noted that because of the limited data available from some of the publications as well as the field experiences, not all analyses were performed by each study. For example, only a few papers examined "regrowth" by storing the cake samples for an extended period of time, and in some cases it is difficult to distinguish between the SI and regrowth. In addition, most investigators did not perform nor did they provide enough data for statistical analyses to be performed. With these limitations in mind, evaluation of the data still provides useful insights. A summary of the process type combinations and whether they reported SI and/or regrowth is shown in Table 3-11. Both SI and regrowth were lumped together due to the limitations of the literature data, but experiences from this research has shown that in most cases where SI occurs, regrowth also occurs.

Based on the results for processes that utilize belt filter presses for dewatering, one out of nine (or about 11%) of these samples exhibited SI and/or regrowth. In comparison, 30 out of 35 plants that utilize centrifuge dewatering, both high and low solids, reported SI and/or regrowth. This is about 86% of the sites with centrifuges. This analysis strongly supports the idea that dewatering processes are important to the SI/regrowth phenomenon. The relatively high percentage of plants with centrifuges that exhibit SI and/or regrowth may be due to some bias in the sampling. For example, the literature likely focused on plants that had this occurring, and therefore, sampling may have been biased toward plants with SI and/or regrowth. Also, since some of the studies did not examine regrowth during cake storage, the percentage of plants may actually be higher. Results from this research have shown that in some cases for mesophilic digestion, SI does not occur, however, regrowth can occur.

Several of the studies performed side-by-side comparisons between belt filter presses and centrifuges that dewatered the same digested product. In three of these cases, the belt filter press cake did not exhibit SI and/or regrowth, however, the centrifuge dewatered cake did.

Table 3-10. Summary of Process that Exhibit SI and/or Regrowth.

Digestion Process	Dewatering Process	SI and/or Regrowth Observed (# Positive/Total Number)
Undigested	High Solids Centrifuge	0/1
ATAD	Belt Filter Press	0/1
MAD	Belt Filter Press	1/6
Pre-Pasteurization + MAD	Belt Filter Press	0/1
Total for Belt Filter Press		**1/9**
ATAD	High Solids Centrifuge	1/1
MAD	High Solids Centrifuge	19/21
Pre-Pasteurization + MAD	High Solids Centrifuge	2/2
Thermophilic	High Solids Centrifuge	4/6
TPAD	High Solids Centrifuge	1/2
MAD	Low Solids Centrifuge	3/3
Total for Centrifuges		**30/35**

Analysis of the plant digestion processes parameters such as SRTs, temperature, and VS reduction was performed to determine if they correlate to the SI/regrowth phenomenon. Digester SRT and VS destruction did not correlate well with the parameters of SI. Figure 3-30 shows the relationship between the digestion temperature and the change in the FC density measured after dewatering. The results show a good, positive correlation between the two parameters. As the temperature of digestion increases, the change in the FC density after dewatering also increases. The results are consistent with previous conclusions, mainly that mesophilic plants typically have low levels of SI, about 0-1 order of magnitude increase, whereas thermophilic processes can have greater amounts of SI. The one 55°C process that appears to be an outlier is the process that utilized reactors in series and did not have SI and regrowth. Also, the 66°C data point is from a pre-pasteurization process. The results seem counterintuitive, in that better destruction of indicator bacteria would be expected at higher temperatures. However, at the higher temperatures, more of the bacteria appear to enter the non-culturable state, and are not enumerated by the SCM. However, qPCR results show fairly high numbers, often around 10^5 per gram dry solids. Therefore, these higher temperatures result in a greater "SI potential" or difference between qPCR and SCM density. The correlation between SI potential and the digester temperature is shown in Figure 3-31, which shows a strong correlation. In other words, at higher digestion temperatures, a greater SI or reactivation potential exists.

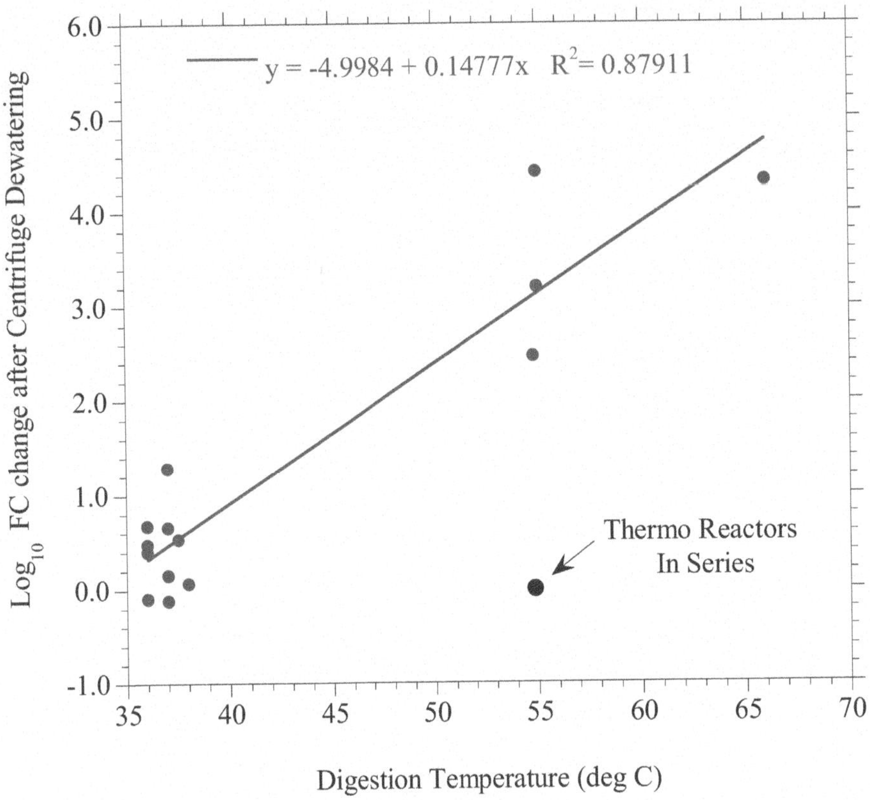

Figure 3-30. Relationship Between Digestion Temperature and Log_{10} Change in FC Density.

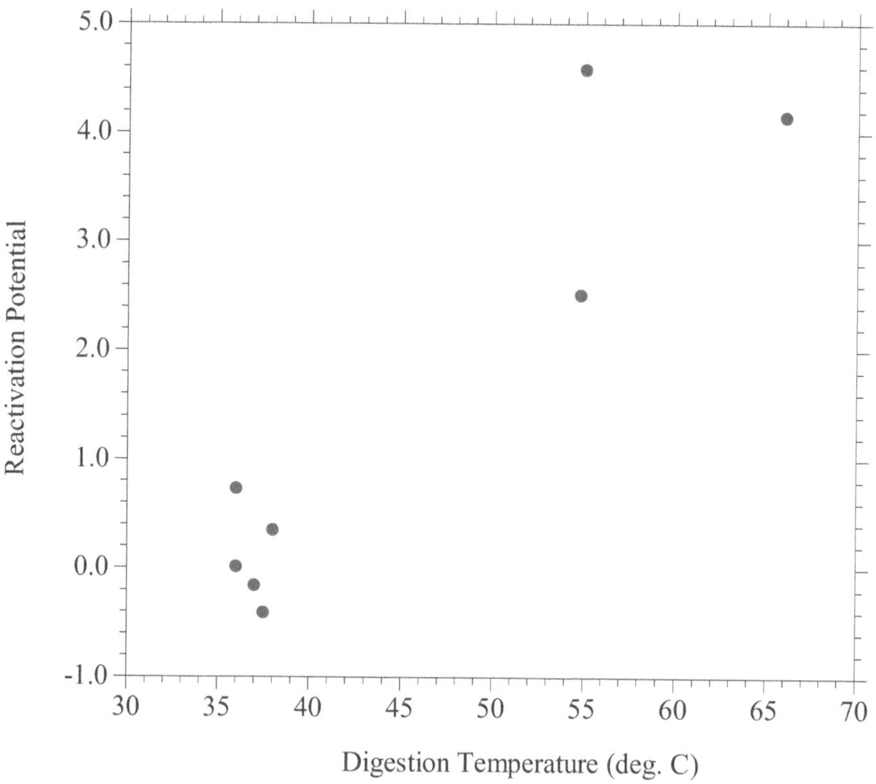

Figure 3-31. Relationship Between Digestion Temperature and SI/Reactivation Potential. (Calculated as the Difference Between qPCR and SCM Measurements of *E. coli*).

Figure 3-32 shows the correlation between the SI potential, and the amount of SI that actually occurred. These parameters show a very strong linear relationship with a correlation coefficient, or r^2 equal to 0.99. Since the SI potential is well correlated to the actual SI, the results are supporting evidence that the qPCR are representative, at least in part, of potentially viable bacteria that are reversibly non-culturable. The slope of the line is 0.80, meaning about 80% of the SI potential is met in most of the samples.

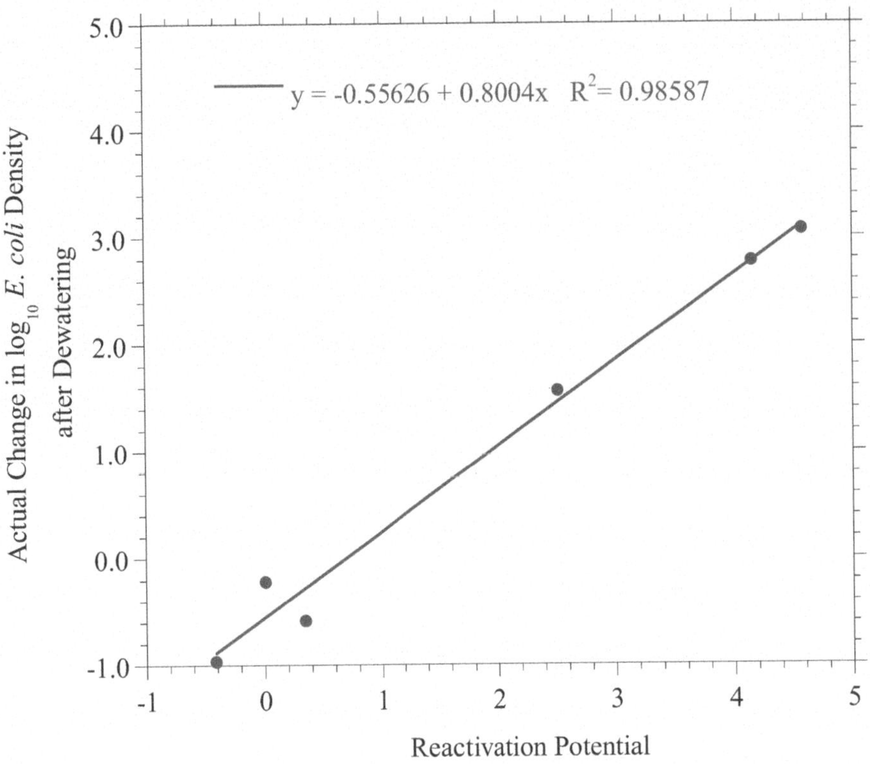

Figure 3-32. Correlation Between SI/Reactivation Potential and Actual Change in the $\log_{10}$ *E. coli* Change After Dewatering.

The regrowth of indicator bacteria during storage did not correlate with digestion SRT, temperature or VS reduction. For example, the change in FC during storage as well as the actual density measured during storage is plotted as a function of digestion temperature in Figure 3-33. During cake storage, the thermophilic processes typically had an additional increase between 2 and 3.5 $\log_{10}$ in the FC density, whereas the mesophilic processes had an increase in FC density between 0.5 and 2.5 $\log_{10}$. The greater increase in FC measured during cake storage for thermophilic processes may be a result of less competition in the thermophilic processes. Analysis of the samples using denaturing gradient gel electrophoresis has shown that the microbial communities are less diverse in thermophilic processes compared to mesophilic processes. With less competition in thermophilic systems, the indicator bacteria may be able to better compete for available substrate, and therefore have, overall, greater regrowth during cake storage. The relationship between VS reduction and regrowth is shown in Figure 3-34, and interestingly shows a weak, but positive correlation. No correlation was found with the digester SRT.

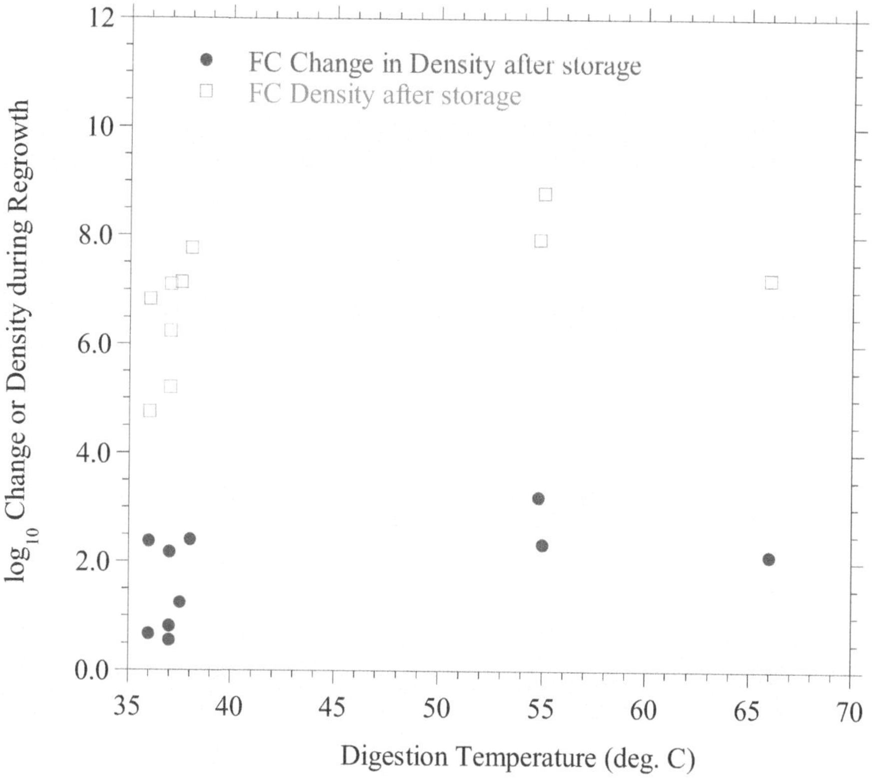

Figure 3-33. Effect of Digestion Temperature and the Amount of Regrowth Measured During Cake Storage.

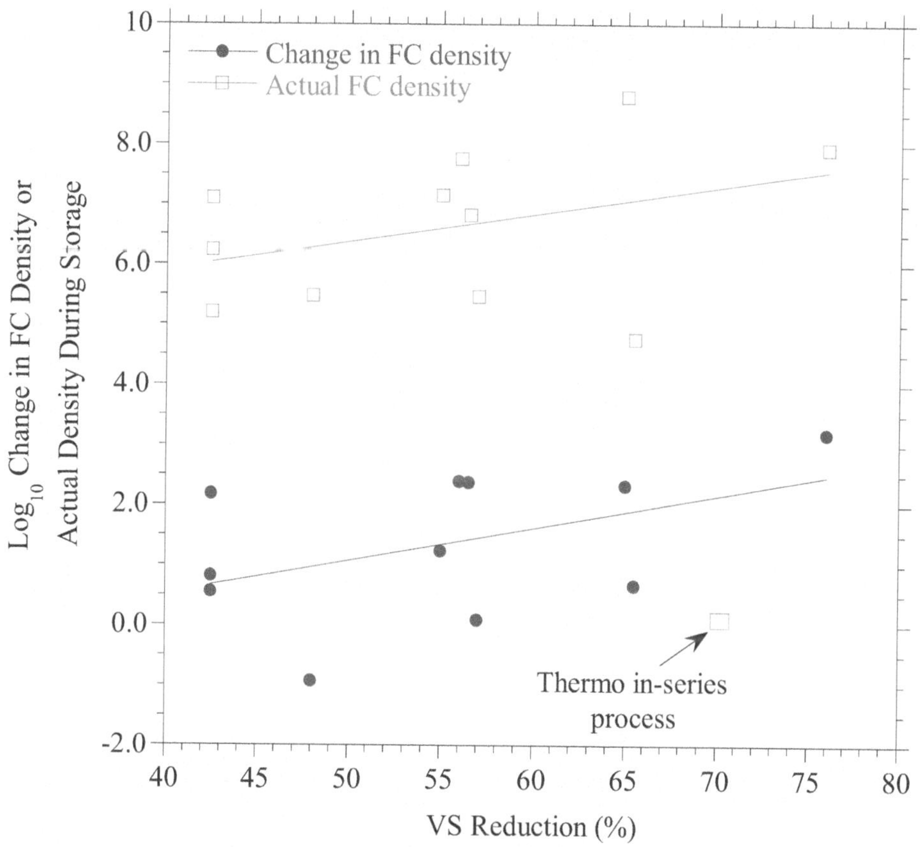

Figure 3-34. Effect of Digestion Temperature and the Amount of Regrowth Measured During Cake Storage.

Several of the processes that were sampled in the WERF studies as well as from the literature are designed using the Class A time and temperatures stipulated by EPA. The U.S. EPA curve is based on literature of pathogen destruction as well as research compiled by Feachem et al. (1983). The U.S. EPA time-temperature curve is embedded within the "Zone of Safety" described by Feachem as shown in Figure 3-35. The graph also shows the different organisms that define the Zone of Safety such as enteric viruses, *Salmonella*, and Ascaris. According to the U.S. EPA requirements, batch digestion using this time and temperature curve, should yield Class A biosolids with FC densities less than 1000/g DS. Using this plot, the different time-temperature combinations used by facilities for which data is available are plotted on the graph in Figure 3-36. The data includes sites from this study as well as the literature. Interestingly, most of the sites are along this curve, yet there are mixed in results in terms of achieving Class A requirements of less than 1000 MPN/g DS of FC. Of the three data points that meet Class A, one utilizes a belt press (all others use centrifuges), one is the thermophilic reactors in series, and the data from Iranpour that did not show SI was after insulating the dewatering train.

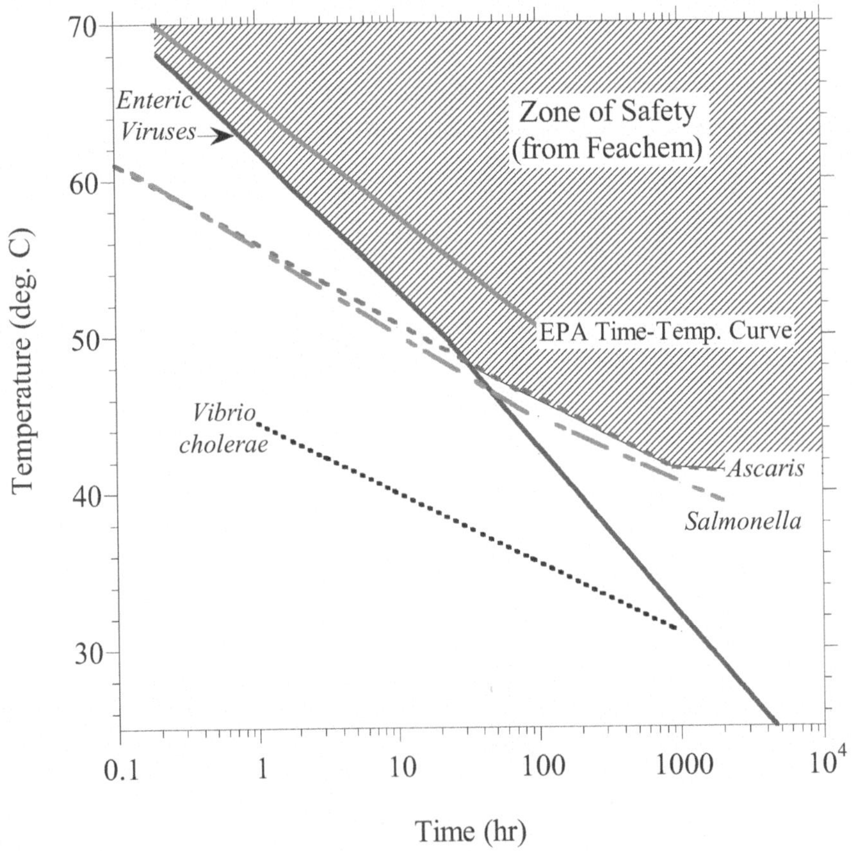

Figure 3-35. U.S. EPA Time-Temperature Curve for Class Digestion and Feachem et al. (1983) Zone of Safety for Pathogen Destruction.

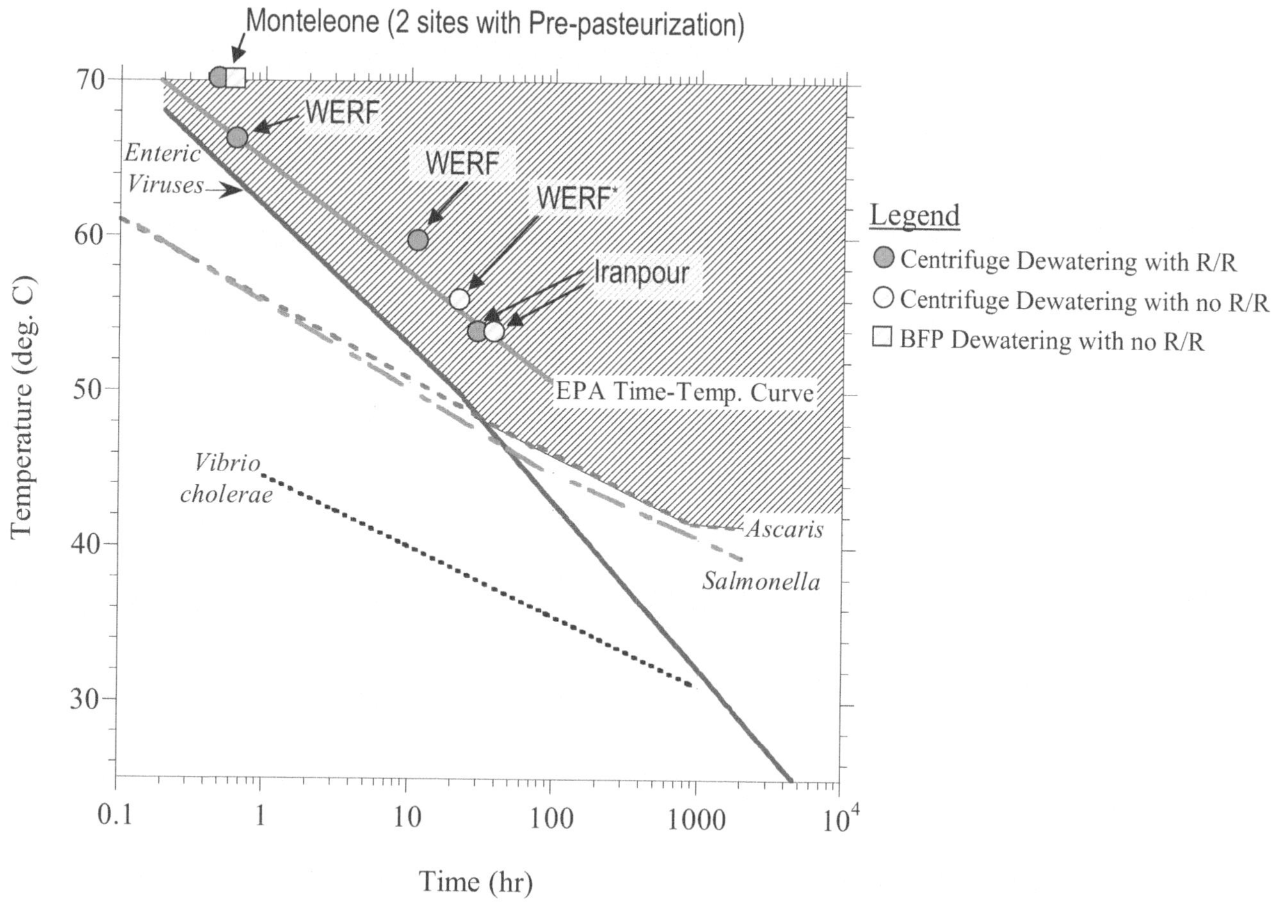

Figure 3-36. Full-Scale Results Compared to U.S. EPA Class A Time-Temperature. Squares Symbolize Belt Filter Press Dewatering; Circles Symbolize Centrifuge Dewatering, Red or Solid Filled Shapes Represent Site with SI/Regrowth; Green, Dashed Shapes Symbolize No SI/Regrowth.

CHAPTER 4.0

RESULTS: TIME-TEMPERATURE EXPERIMENTS

4.1 Effect of Mesophilic SRT on VBNC *E. coli*

A study was conducted to investigate the effect of digester SRT on the density of *E. coli* as measured by qPCR and the SCM. In this study, laboratory scale completely mixed, mesophilic (35°C) digesters were operated for a minimum of two or three SRTs to achieve steady state. The SRTs that were studied were 10, 15, 20, 30, and 40 days. After steady state conditions were achieved, samples were analyzed for *E. coli* using the MPN culturing procedure and using qPCR. A comparison of the *E. coli* densities measured by the two methods is shown in Figure 4-1.

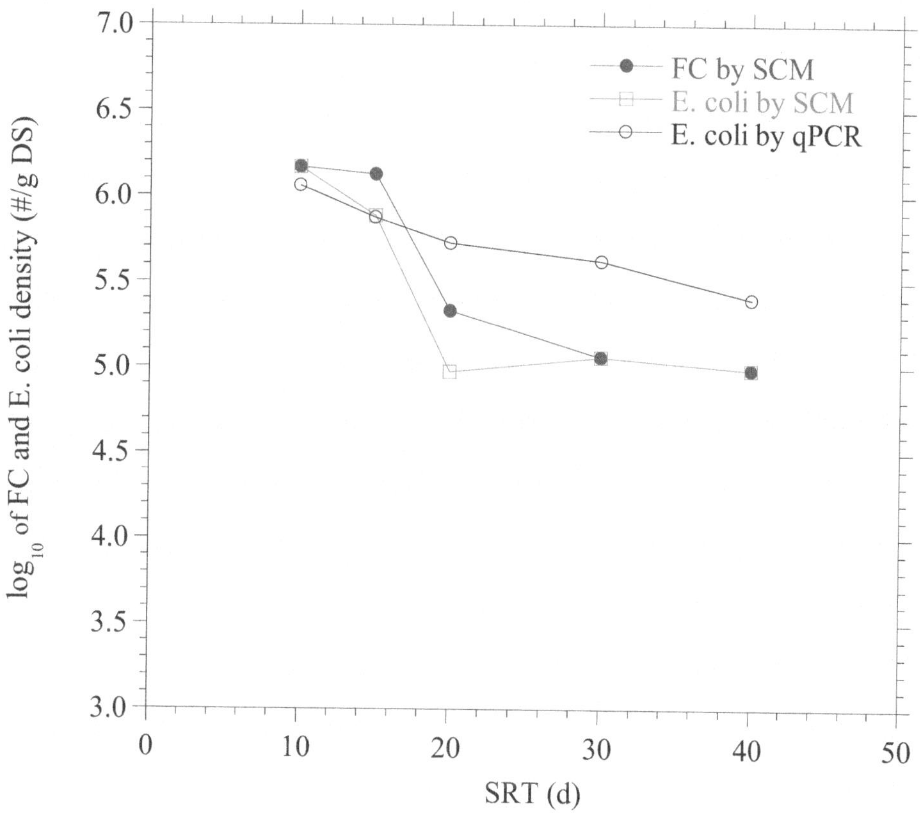

Figure 4-1 *E. coli* Densities as a Function of SRT as Measured by qPCR and an MPN Method.

From previous testing, non-culturable *E. coli* can be considered as the difference between qPCR density and the SCM density. Examining Figure 4-1, the difference between qPCR and SCM increases at SRTs greater than 15 days, and the maximum difference is around 20 days of digestion. After this period, the SCM results are fairly constant, and the qPCR results slowly decrease. It is interesting to note that no improvement occurs after 20 days based on the MPN method, and the densities remain near 10^5 per gram DS. Results from the field in Mesophilic Plant 1 achieved an effluent density between 10^5 and 10^6 cells/g DS, with an SRT of about 31 days. The abrupt decrease in log FC reduction by the MPN method on day 20 may be an anomaly. Other investigators show smoother changes in FC as SRT increases (Lucero-Ramirez, 2000; Cacho Rivero, 2005).

The results from Phase I of this research showed that digester effluent generally had indicator densities between 10^5 and 10^6 *E. coli*/g DS, as measured by qPCR, suggesting that mesophilic digestion can typically meet Class B requirements (assuming *E. coli* are the main FC organism). The numbers of non-culturable *E. coli* were much less in mesophilic digesters compared to thermophilic digesters. In other words, *E. coli* remain largely culturable after mesophilic digestion. As a result, the increase of the indicator organisms after high solids centrifugation is in part due to SI of some non-culturable *E. coli*, but more so due to their growth in the cake. The ability to grow in the cake suggests that the environmental conditions may have changed to promote the growth. Research has shown that the shear associated with high solids centrifugation may release substrates that can become available for bacterial growth (Higgins et al., 2006; Murthy et al., 2003).

4.2 Time-Temperature Effects at Thermophilic Temperatures

Samples were collected from laboratory scale anaerobic digesters operated by Virginia Tech. These were batch fed reactors, operated at an SRT of 15 days, with different temperatures of 35°, 49°, 53°, and 57.5°C. For more details on the operation of the reactors, please see Chapter 2.0, Methods and Materials. The reactors were sampled, and analysis of FC and *E. coli* was performed using both SCM and qPCR. A sample of the feed solids was also analyzed at the same time. The mean density of FC and *E. coli* measured by both methods is shown in Figure 4-2. The digester feed, which was a mix of primary and secondary sludges, had an FC and *E. coli* density in the range of 10^7 cells/g DS measured by the two methods which is typical of most undigested sludges sampled in this research and others. The results from the mesophilic digestion are also very similar to the field observations from this research. The qPCR values were about 0.6 $\log_{10}$ greater than the SCM value. For many of the field samples, the difference was in the range of 0 – 1 $\log_{10}$ difference. For the higher temperature reactors, the difference was much greater between the two methods. The SCMs measured about 10^1 or less FC and *E. coli* densities. It should be noted that for the triplicate analyses using SCMs, several were below the detection limit for the dilution set used (about 8-10 MPN/g DS), and a value of half the detection limit was used for calculating the geometric mean. In the case of the 57.5°C reactor, the values were all less than 8 MPN/g DS, and a value of 4 MPN/g DS was used for graphing purposes. The SCM densities match well with field samples utilizing thermophilic digestion which were typically less than 100 MPN/g DS. The qPCR method measured between 10^4 and 10^5 *E. coli* per gram of dry solids. The qPCR values also match well with results from the two full-scale single stage thermophilic digestion sample analyses performed as part of this study. The $\log_{10}$ difference between the qPCR and the SCM was 3.3, 4.7, and 4.6 for the temperatures of 49°, 53°, and 57.5°C, respectively.

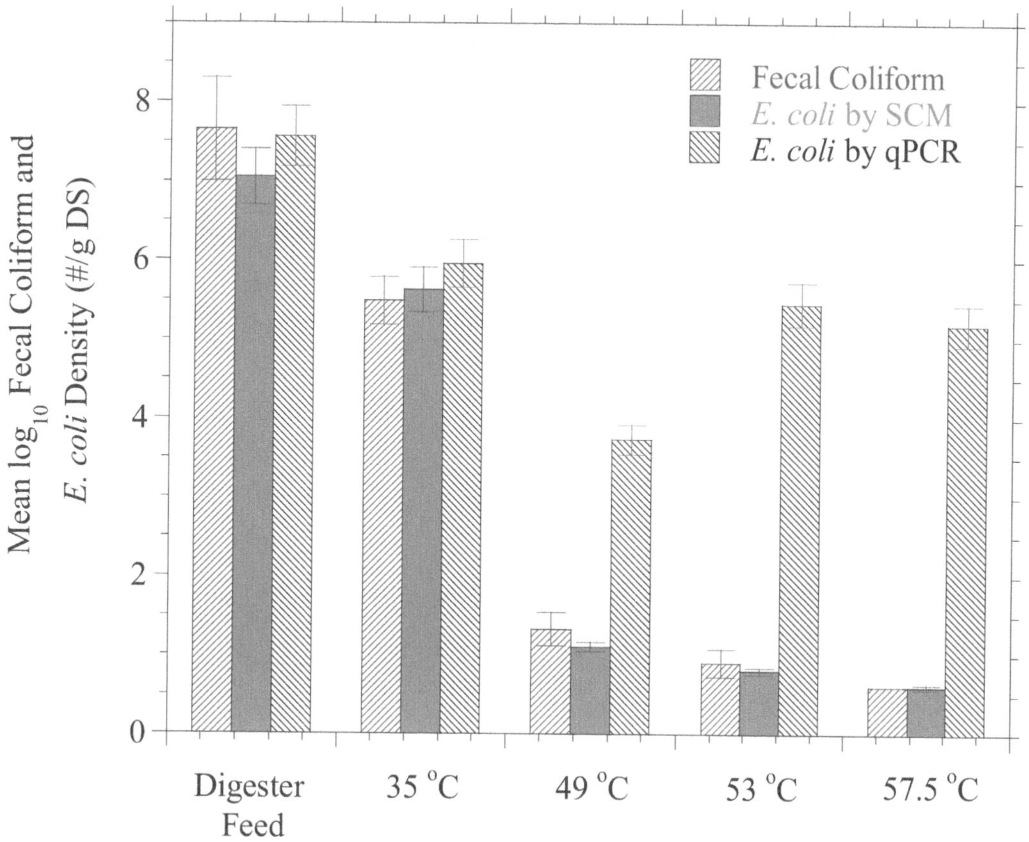

Figure 4-2. FC and *E. coli* Densities for Different Reactor Temperatures as Measured by qPCR and MPN Method.

Since the batch digestion time was approximately 12 hours after feeding for this first trial, a second longer term study was performed to examine the impact of long term batch digestion on the density of indicator organism measured by both SCM and qPCR. In this study, the reactors were operated until steady state conditions were achieved using daily feedings, then at the beginning of the study, the reactors were fed, and then samples were taken at regular intervals and analyzed for FC by the SCM and *E. coli* by both the SCM and qPCR for about 60 days. During this time, the reactors were not fed. The density of *E. coli* measured by SCMs during the first 24 hours is shown in Figure 4-3. The FC results were very similar to *E. coli* (data not shown). The decrease in culturable *E. coli* occurred fairly rapidly during this period as would be expected from previous research. However, the *E. coli* density measured by qPCR did not decline as rapidly, as shown in Figure 4-4. The densities decreased relatively slowly over the first 30 days, and then reached a steady density around 10^5 cells/g DS through about 60 days of measurement for each temperature. The densities measured through about 20-30 days of batch digestion are similar to those from the field thermophilic single stage digesters with about 10^5 *E. coli* per gram of dry solids.

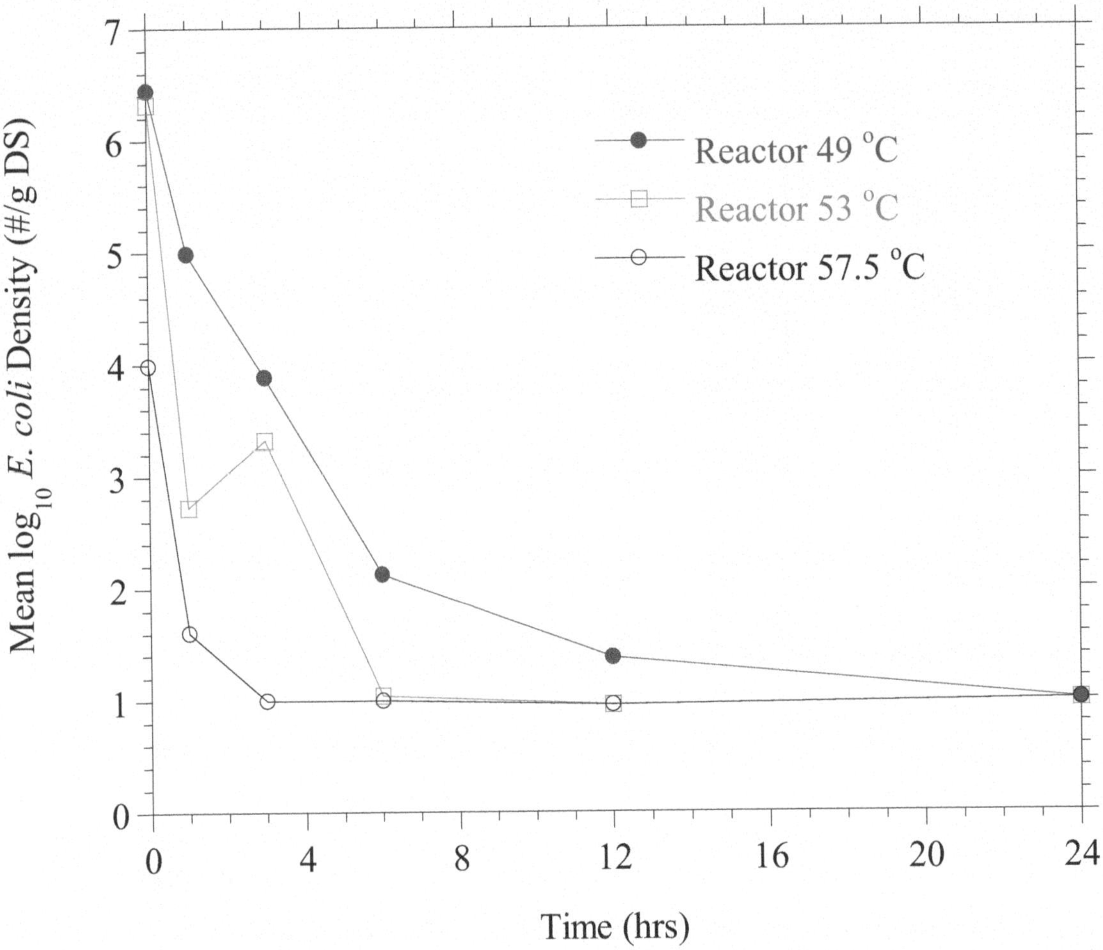

Figure 4-3. ***E. coli*** **Densities Measured by SCM During Batch Digestion at Different Temperatures.**

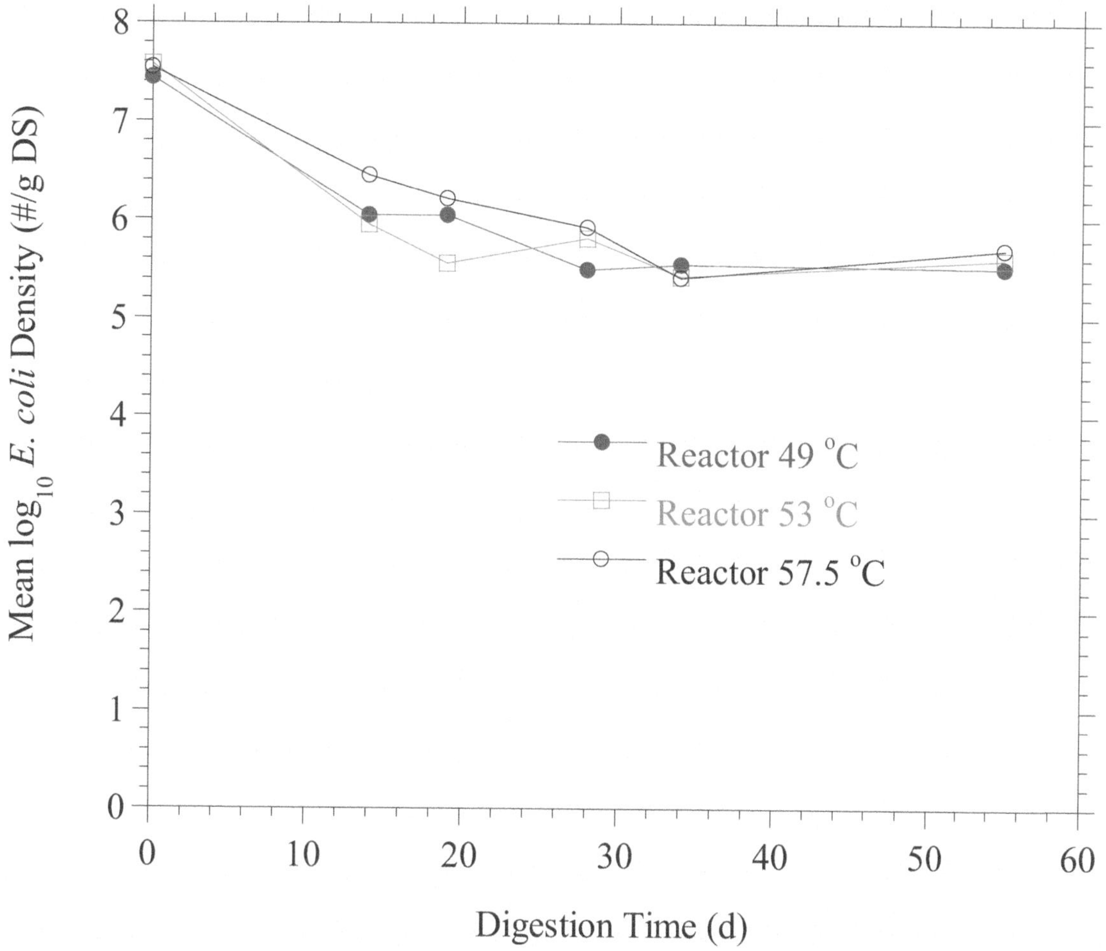

Figure 4-4. *E. coli* Densities Measured by qPCR During Batch Digestion at Different Temperatures.

Beyond the time period after about 30 days, the results are somewhat surprising in that conventional practice would not have expected these densities to remain for such a long period. Several possible explanations exist for these qPCR results:

1. the data are erroneous;
2. the *E. coli* DNA measured is due to undegraded, free DNA in solution from lysed cells;
3. the *E. coli* DNA measured is due to dead or nonviable *E. coli* that have not yet lysed, and, as a result the DNA is undegraded;
4. the *E. coli* DNA measured is due to VBNC *E. coli* that are still viable and remain;
5. some combination of two or more of the above possibilities.

Each possibility is considered in the following paragraphs.

It seems unlikely that the data is in error for several reasons. First, the decay in *E. coli* DNA during batch digestion proceeds relatively smoothly, with little deviation or scatter in the data. In addition, a number of quality assurance measures are used during the qPCR analysis to reduce the possibility of erroneous data, so we have confidence in the results.

Second, the *E. coli* DNA that was measured could result from free, undegraded DNA from dead cells that have gone through cell lysis. The released DNA could then persist in solution, and then be measured during extraction and qPCR analysis. However, several studies have shown that free DNA is rapidly degraded in wastewater and environmental samples with active microbial communities (Ruiz et al., 2000; Lorenz and Wackernagel, 1994; Fibi et al., 1991; Philips et al., 1989). Although, researchers have also shown that when DNA is bound to surfaces such as clay or minerals, the degradation by nuclease enzymes is reduced (Lorenz and Wackernagel, 1987; Demaneche et al., 2001). Therefore, it is possible that DNA released from lysed cells binds to surfaces available in the sludge, and DNA degradation is then potentially inhibited. Sludges provide numerous surfaces for binding opportunities. In addition, biological activity as observed by qualitative gas measurement showed that gas production in the reactors decreased during the study to a very low amount around day 30. Therefore, the low biological activity may also contribute to the persistence of free DNA in the sample.

The third possibility is that the *E. coli* DNA enumerated in the samples is from dead or non-viable cells that have not yet lysed, and remain intact, and therefore their DNA remains intact and not degraded. Examination of the results in Figure 4-4 shows that the *E. coli* density decreased steadily for the first thirty days, therefore, assuming some *E. coli* are dying, their cell DNA is being consumed and the densities decrease. This indicates that, indeed, DNA can be degraded in anaerobic digestion as others have reported (Ruiz et al., 2000), which would presumably occur after cell lysis. However, it is not clear why the decrease in *E. coli* density does not continue between the time period of about 25-60 days. One possibility is that the biological activity had decreased to the level below which significant degradation would no longer occur. Qualitative assessment of gas production in the reactors has shown that gas production decreased to almost no gas production some time around 30 days of storage, which might help explain these results. However, it would seem that if available DNA existed, it would be degraded as it supported cryptic growth of the microbial population.

The fourth possibility is that the *E. coli* DNA measured in the samples represents reversibly, non-culturable bacteria, such as sublethally injured or VBNC *E. coli* that are still potentially viable, and capable of being reactivated. For example, studies have shown that VBNC bacteria can survive for prolonged periods, and in fact, the VBNC state is a survival mechanism of bacteria that allows them to persist under stressful conditions such as starvation. An example of this is the change in physiology that has been observed in VBNC bacteria (Signoretto et al., 2000; Heim et al., 2002). Signoretto et al. (2000) reported that VBNC *Enterococcus faecalis* had cell wall peptidoglycan with greater cross-linking than normal growing cells and this provided greater resistance to mechanical disruption. A key component of the VBNC state is the ability of the organisms to be resuscitated or reactivated, which is typically defined as the ability to cause infection *in vivo* or to grow on amended media or in the case of this research to grow after a "reactivation event" such as has been observed to occur with high solids centrifugation.

It is not clear from the data which of these possibilities are most likely, and it is not clear how long *E. coli* can survive at different temperatures during anaerobic digestion, mainly because the field does not have a reliable measurement of viability nor a reliable method for SI or resuscitation of the indicator organisms. In addition, these results further highlight the need for developing a laboratory scale approach for reactivating non-culturable organisms. This research team has undertaken some preliminary experiments to determine if SI can be induced in the laboratory, and these results are presented in Chapter 5.0.

To gain some insight into these possibilities, an experiment was performed to determine if "naked" DNA, or DNA that has entered the extracellular environment, is degraded during anaerobic digestion, and especially examine the effect of biological activity on naked DNA degradation. This data could then be used to help explain, in part, the results from Figure 4-4 which showed stable DNA concentrations for an extended period. A thermophilically digested sludge from the 53°C reactor was stored for 90 days, which exhibited no gas production and hence low biological activity, was used to examine the effect of biological activity on DNA degradation. *Eschericia blattae* DNA was used for the test since it was found to not exist in measurable quantities in digested sludge (data not shown). The cells were grown in batch culture, and the DNA was extracted and quantified. A specific amount of *E. blattae* DNA was then spiked into sacrifice bottles containing digested sludge, with and without additional COD added to the bottles. The COD was added to achieve a final concentration of 10 g/L from glucose and bactopeptone at a 9:1 mass ratio, respectively. A real-time PCR primer set was designed and used to quantify *E. blattae*.

The density of *E. blattae* DNA copy numbers for the different conditions is shown in Figure 4-5. The results showed that after spiking, a reduction of two orders of magnitude in the DNA density resulted. This was likely due to centrifuging of the samples to concentrate the solids, followed by removal of the supernatant that would contain potentially naked DNA which is the standard method used as part of this research. Therefore, this suggests that the lab centrifugation step is important, and removes approximately 94-96% of the naked DNA, and the remaining fraction is bound to the solids. The results also indicated that the addition of substrate to stimulate microbial activity was important in the degradation of the naked DNA. For example, the density of DNA copies decreased by three orders of magnitude during approximately the first six days of storage in the sample set amended with substrate to stimulate biological activity. In comparison, the sample set without substrate had about a 0.5 $\log_{10}$ reduction in DNA copies. However, the results also implied that a possibility of inclusion of naked DNA in enumeration is likely if the sample is stable, i.e. biologically inactive. Overall, naked DNA is readily degraded in a biologically active digester, but slowly in a stable digester.

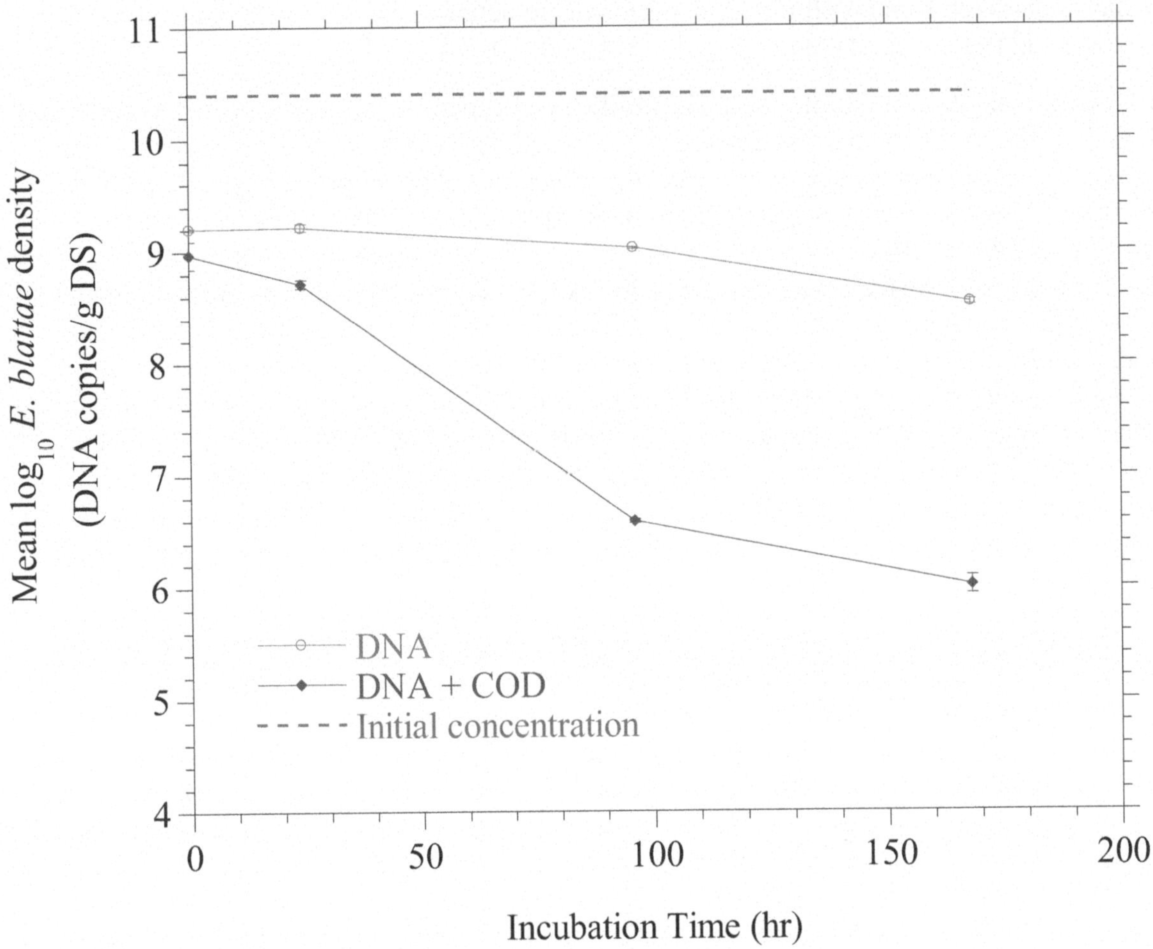

Figure 4-5. Densities of *E. blattae* Naked DNA Measured by qPCR During Batch Digestion at 53°C, with and without Addition of Substrate.

CHAPTER 5.0

LABORATORY EXPERIMENTS

5.1 Introduction

Developing a resuscitation method to better enumerate VBNC organisms is of critical importance to this research. The method will allow determination of viable organisms that were previously not culturable, and will open the doors to many avenues of key research needed to fully understand the SI and regrowth issue. For example, the large scale field experiment at Thermophilic Plant 2 was undertaken in an attempt to determine if SI would occur after two-stage digestion. In order to do this, full-scale dewatering using a high solids centrifuge was needed, and the plant personnel had to expend significant effort to allow dewatering of the first, second, and third stage reactors directly to the high solids centrifuge. Unfortunately, due to certain site limitations, at least two reactors in series was the minimum number that could be examined in the study. As discussed in the results in Chapter 3.0, the field trial was performed, however, due to several problems, the data set was not complete, and the goals of the study were not completely achieved.

If, instead, a laboratory-scale approach existed that resuscitated or reactivated the FC and *E. coli*, samples from each digester could have been obtained and run through this procedure. This would have significantly reduced the effort required to perform the trial, and the limitations associated with full-scale work would have been eliminated. Similarly, a lab scale method could be used in the time-temperature experiments to better understand at what point the *E. coli* might become nonviable.

Several preliminary experiments have been performed to determine the potential of these approaches to either resuscitate non-culturable bacteria using media additives, or to reactivate the bacteria using a process to simulate a full-scale centrifuge.

5.2 Resuscitation with Media Additives

A number of researchers have shown that the selective media used to enumerate FC and *E. coli* while preventing growth of other microbes can underestimate the numbers of target organisms that are actually enumerated (Martin et al., 1976; McFeters et al, 1982). This has been shown for both the sub-lethal injury and the VBNC theories, which are similar theories. It is typically thought that the organisms under stress are less able to cope with free radicals or oxidative stress such as the presence of hydrogen peroxide, which limits their ability to grow. Addition of scavengers such as pyruvate or catalase to the media has been used to reduce the oxidative stress and increase the counts with selective media (Marin et al., 1976; McDonald et al., 1983; Mizunoe et al., 1999). This result makes the term "viable but non-culturable" somewhat of an oxymoron, because the bacteria are in fact culturable, just not under the standard conditions used to enumerate the organisms.

Another possible explanation for the lack of growth using standard culturing methods could be that growth inhibitors are formed during digestion, and inoculation into the media as well as dilution is not enough to remove these inhibitors and induce growth.

A number of experiments were performed in which additives such as catalase, pyruvate and autoinducers were added to the media according to the methods used in the literature to determine if higher densities would be enumerated. In addition, several washing experiments were performed in an attempt to remove possible inhibitors that might be present. These experiments were performed using a digester effluent from a TPAD process as well as effluent from a mesophilic digester (Mesophilic Plant 3).

Several of the experiments produced statistically significant differences, but many did not. The results from the experiment that utilized washing of the sludge as well as addition of different autoinducers and the autoinducer stimulator norepinephrine are shown in Figure 5-1. The only treatment that had a significant effect in this trial was washing, which may indicate the presence of some inhibitors, but the increase was for total coliform (TC) density and only about one order of magnitude for the TPAD sample. No increase in FC or *E. coli* was measured in the samples.

In a second set of experiments on a separate date, several hydrogen peroxide scavengers were added to the media which included catalase, pyruvate, and alpha-ketoglutarate. The addition of these compounds has been used successfully by other investigators to increase the numbers of sub-lethally or VBNC bacteria enumerated by culturing methods. The results from these tests are shown in Figure 5-2. The only treatment that had a statistically significant effect was pyruvate addition, which increased the FC density measured by about one order of magnitude compared to the control. A third trial was performed using anaerobically digested solids from a plant with pre-pasteurization. In this test, addition of anti-oxidants was tested. An increase in total coliforms was measured with the control, but no FC or *E. coli* were enumerated in any of the media types (data not shown). These initial results are mixed, but show that resuscitation may be possible by media enhancements. Much of the literature on this topic focused on mono-cultures which makes the analysis much simpler compared to biosolids which are a complex consortium of microbes, organic matter, nutrients, and inorganic matter. This complexity may make resuscitation more difficult in biosolids compared to other media, and a significant amount of additional research is needed to optimize the culturing media and growth conditions to maximize the enumeration of previously non-culturable bacteria, and to verify the results for a number of plants.

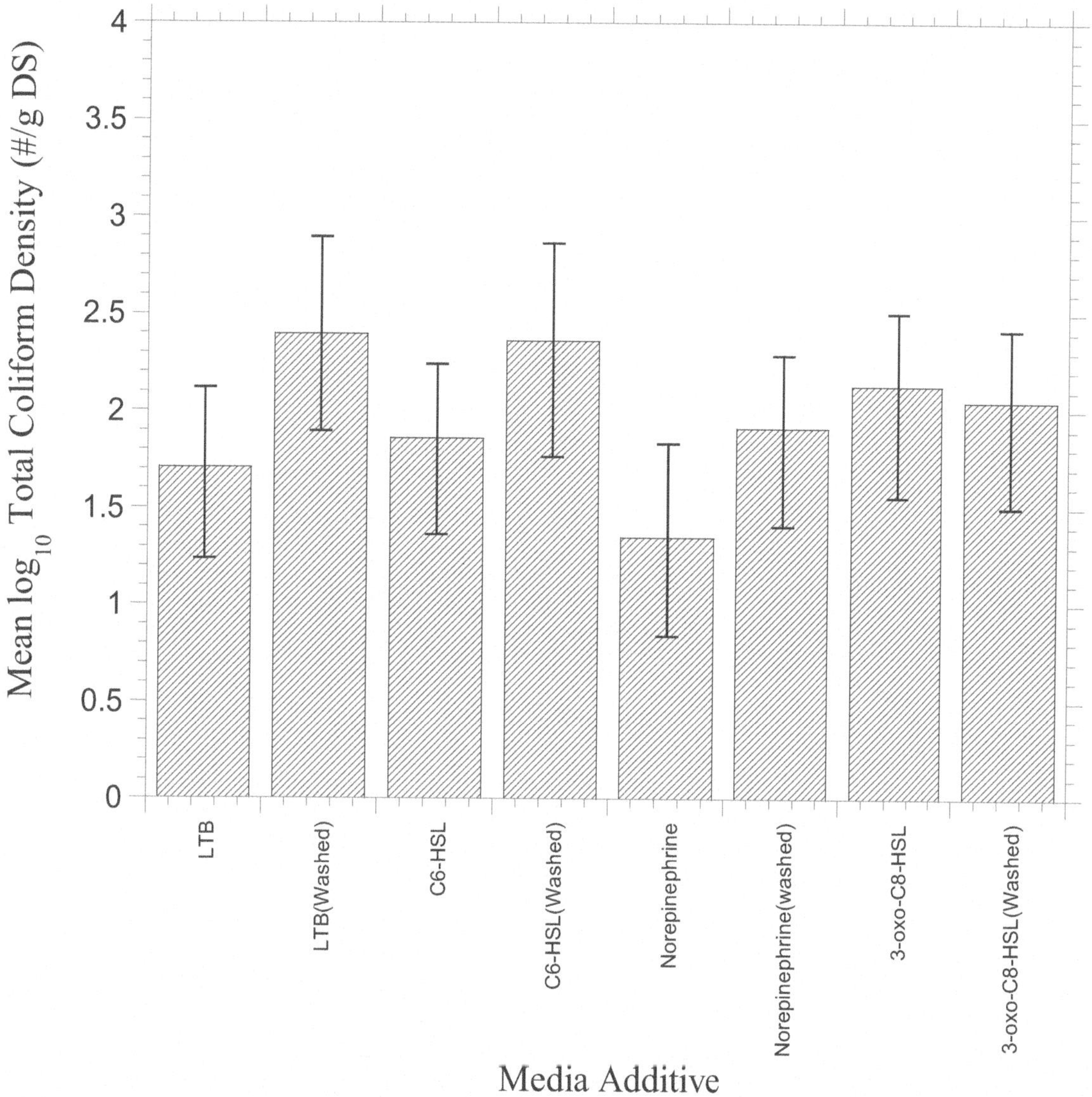

Figure 5-1. Effect of Autoinducer Addition and/or Cell Washing on Resuscitation of Total Coliforms.

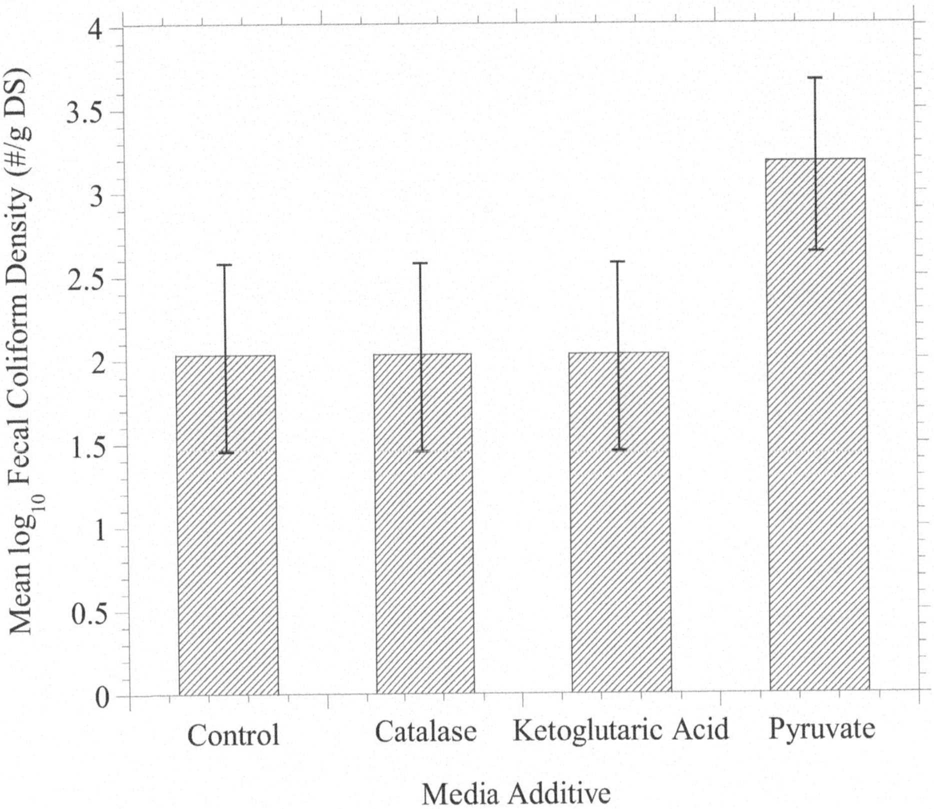

Figure 5-2. Effect of Oxygen Radical Scavenging Media Additives on Resuscitation of Fecal Coliforms.

5.3 SI with Centrifuge Simulation

Another possible method to enumerate the nonculturable fraction of bacteria is to simulate the shear associated with a high solids centrifuge. This method could also have the potential to allow the study of regrowth since centrifuge cake also supports regrowth by providing substrate and nutrients. The mechanism by which high solids centrifugation causes SI is thought to be due to shear imparted to the sludge as it is dewatered. Higgins et al. (2006) showed that, in general, high solids centrifugation had a greater shear than belt filter press dewatering. This shear may cause SI through the release of inducer-like compounds, or provide a change in environmental conditions which allows the previously nonculturable cells to become culturable. The physical component of shear in the high solids centrifuge to make the cells culturable is termed "reactivation" for the purpose of this project, whereas resuscitation has been associated with a change in the media type or culturing conditions to result in the increase in culturable organisms. An advantage of a reactivation method that simulates high solids centrifugation would be that regrowth could also be studied.

Previously, a high solids centrifuge simulation (HSCS) procedure was developed at Bucknell University for the purpose of biosolids odor studies. The procedure produces cakes with similar odor profiles as cake samples from full scale centrifuges. However, no testing has been performed to determine if the method also causes SI and supports regrowth during cake storage. It is interesting to note that the SI and regrowth phenomena observed in many of the samples from high solids centrifuges may be related to odor production during cake storage

(Chen et al., 2007). For example, results have shown that growth of *E. coli* during storage followed a very similar pattern as the production and decay of the main odor causing compounds, total volatile organic sulfur compounds (TVOSCs), as shown in Figure 5-3.

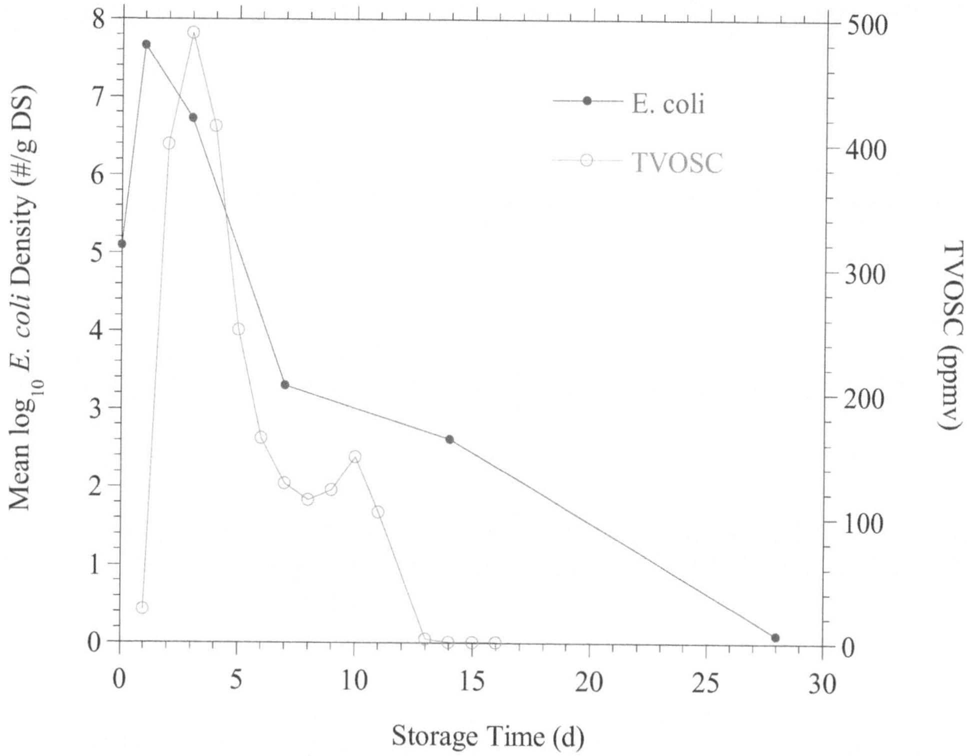

Figure 5-3. ***E. coli*** **Growth and Decay and Production of Odorous TVOSCs During Cake Storage.**

Several experiments were performed in which the HSCS method developed at Bucknell University for the purpose of odor research was tested to determine if it also caused SI and could support regrowth. The HSCS first conditions the liquid sludge with cationic polymer using a relatively high shear mixing regime. The mixing utilizes a Gt value of 100,000, where G is the velocity gradient (in this experiment G=1000/s), and t is the time of mixing in seconds. The unitless parameter of Gt is often used to quantify mixing intensity. After the sludge is conditioned, it is then centrifuged in a laboratory centrifuge at 3000 x g, for 10 minutes. The resulting cake typically has a solids content around 16-19%, and it is then further subject to enhanced belt filter press dewatering to increase the cake solids to between 25-30% which is a typical cake solids range for high solids centrifuges. Finally, the cake is then subject to shear by putting it through a meat grinder apparatus on a KitchenAid mixer. This meat grinder has a scroll that pushes the cake forward, and then the cake is extruded through small openings at the end of the scroll. This shear on the cake is thought to simulate the shear in a high solids centrifuge. Using this method, cakes can be produced that have the same odor generation patterns as the cakes from the same liquid solids produced on full-scale centrifuges. The full method is described in Section 2.5.2.

A sample of digester effluent (centrifuge feed), centrifuge cake, and polymer was collected from Mesophilic Plant 3, and transported to the Bucknell University Environmental

Engineering and Science Laboratory, and subject to the HSCS procedure. Prior to the experiment, all equipment was sterilized in order to prevent contamination. For the purpose of the SI investigation, the cake was passed through the meat grinder with either 0, 5 or 10 times to see how different amounts of shear may impact SI and regrowth. After the HSCS procedure, the samples were immediately analyzed for FC and *E. coli*, and the remaining cake was stored at mesophilic temperatures to examine for regrowth. Samples were taken on days 1, 2, 4, and 7 for the regrowth study. The field cake from the plant's full-scale centrifuge was also analyzed for regrowth for comparative purposes.

The results from this experiment are shown in Figure 5-4. The HSCS procedure was able to produce a regrowth curve very similar to the field cake with 5 or 10 passes through the system. Interestingly, 0 passes through the system did not produce statistically significant regrowth or increase in FC density during storage ($p<0.05$). The field cake, 5, and 10 passes all produced statistically significant increases during storage. The 0 pass cake was only a belt filter press dewatered cake, with no additional shear which is similar to the field belt filter press results that did not produce statistically significant regrowth during cake storage both from this study and others (Erdal et al., 2003; Qi et al., 2004). Lab centrifuging alone has also not been shown to cause SI or regrowth in this study and by Iranpour et al. (2002) and Monteleone et al. (2004). Similarly, several researchers reported that shear of the liquid by a kitchen blender or homogenization did not cause significant SI (Qi et al., 2004; Cheung et al., 2003). The HSCS process uses a relatively high shear during conditioning of the liquid solids, but this was unable to support regrowth. The additional shear associated with the scroll and extrusion was able to support regrowth. Unfortunately, in this study, SI was not found in the field cake nor in the cake from the HSCS procedure. No qPCR samples were taken to determine the potential for SI since this was a preliminary trial.

A second trial was performed. In this trial, the batch stored mesophilic digested sludge from Figure 3-22 was processed through the HSCS procedure after the liquid sludge was stored for 7 days. At this point, the FC density was about 10^2 MPN/g DS. The liquid was conditioned and dewatered as previously described, and then put through the scroll/extrusion process either 5 or 10 times. Immediately after the scrolling/extrusion process, a subset of the cake samples were enumerated for FC and *E. coli*, and the remainder were stored under mesophilic conditions for periodic sampling. The results from this test are shown in Figure 5-5. The sample at time zero did not show any increase in FC immediately after processing. However, after one day of storage, the density increased by 4-5 orders of magnitude for the different shear amounts. The peak densities occurred after the first 1-2 days of storage, and decreased with further storage, similar to field trials. Interestingly, no sudden increase was observed in this test. Enumeration of *E. coli* DNA density by qPCR revealed that SI potential existed, as shown in Figure 5-6. The difference in the culturable and DNA density was about four orders of magnitude. Therefore, the method did not appear to cause SI, but did result in regrowth. However, since the samples were analyzed almost immediately after processing, the time needed for bacteria to reactivate may not have been sufficient. For example, the literature has shown that methods for resuscitation or SI of non-culturable bacteria may take on the order of hours for complete SI although in some cases it is immediate. During field sampling when SI has often been measured, the cakes are placed on ice, and held for a period of hours on the ice until analysis is performed. Although the cold temperatures would prevent any significant growth during this period (something also confirmed in this study), the bacteria could potentially have time to reactivate, such that when placed in the standard culturing media, they grow. This is supported by evidence from a utility in Australia that stated during a WERF presentation on this topic, when samples were taken from the centrifuge and analyzed immediately, the problem of SI disappeared, compared to having a short holding time after collecting the sample. This suggests that although SI was not measured in the lab tests, if a waiting period was employed, the SI may have been observed. Another supporting piece of evidence that SI may have occurred is the doubling times measured in the testing. The field cake used in the first test had a doubling time of about 230 minutes during the first 24 hours. In comparison, the doubling time in the first 24 hours of the *E. coli* data shown in Figure 5-5 is about 120 minutes. This faster rate of increase in the *E. coli* density might be due to a combination of SI and regrowth.

The results from these lab trials are very encouraging in that a laboratory procedure has been investigated that has the potential to be used to study SI and regrowth. Additional research is needed to further refine this method and verify it with side-by-side testing of full-scale centrifuge results. However, the initial results are promising.

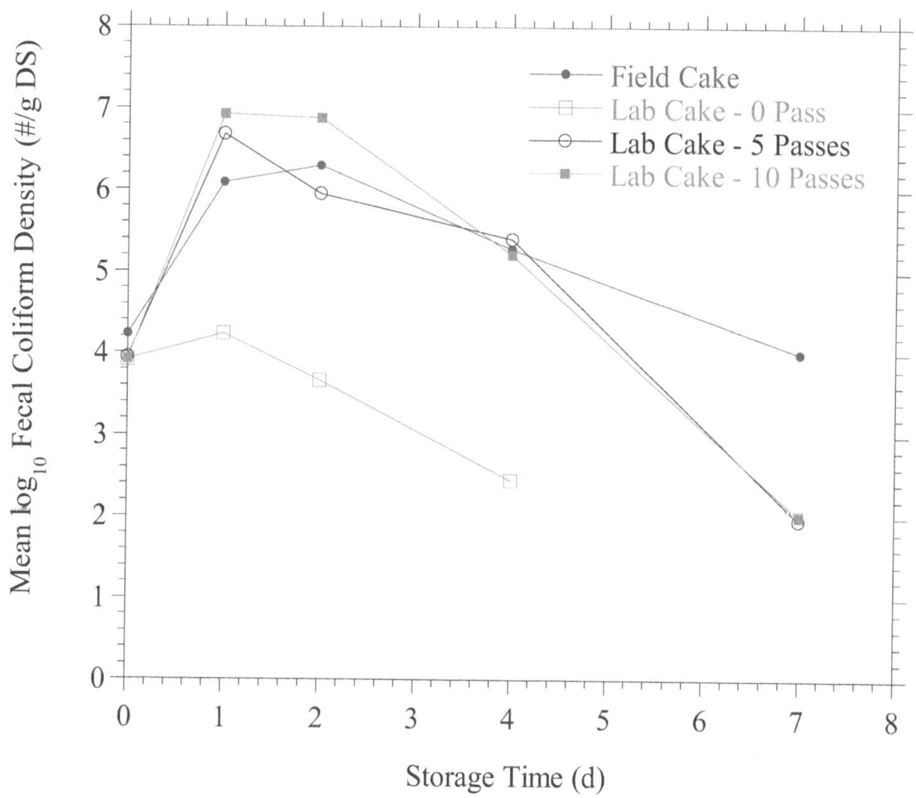

Figure 5-4. FC Growth and Decay During Cake Storage from High Solids Centrifuge Simulation Procedure.

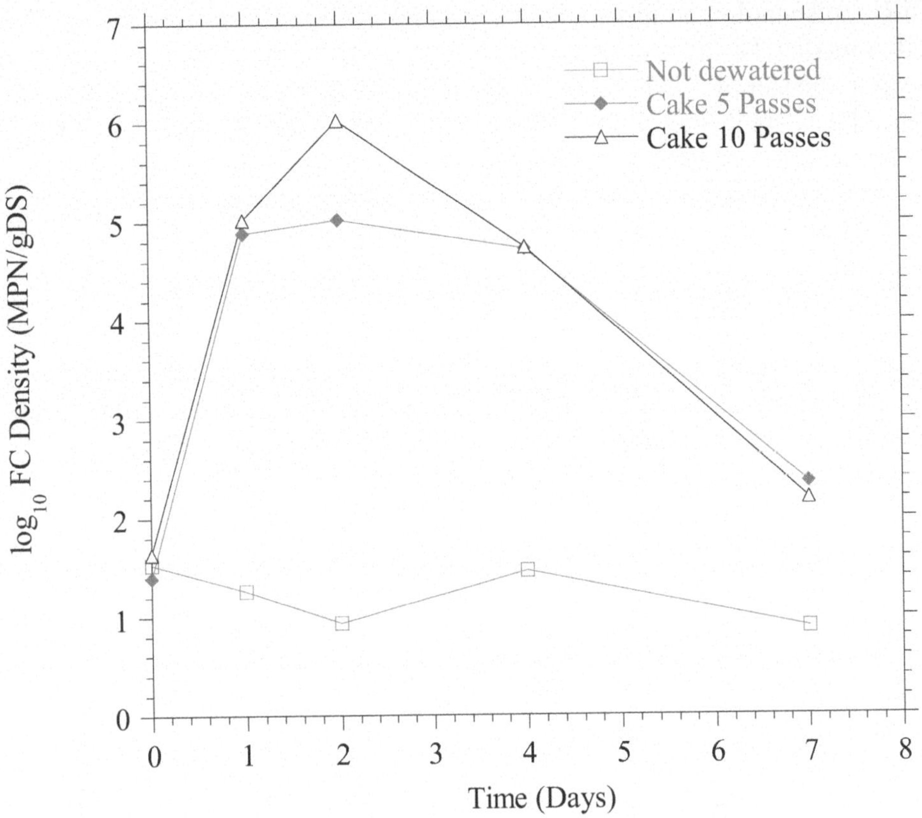

Figure 5-5. FC Growth and Decay During Storage of Cake Produced from High Solids Centrifuge Simulation Procedure, and the Liquid Sludge Used to Generate the Cake, but Not Processed Through the Procedure.

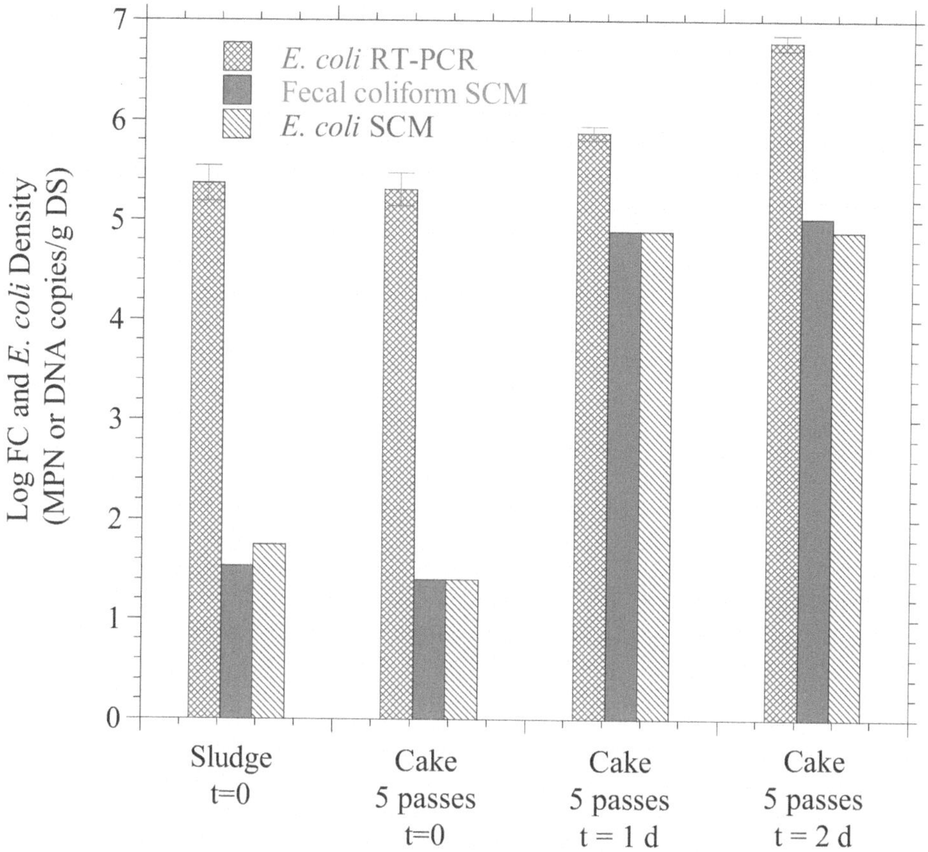

Figure 5-6. Comparison of FC and *E. coli* by SCM and *E. coli* by qPCR During Storage of Cake Produced from High Solids Centrifuge Simulation Procedure with 5 Passes Through the Scroll/Extrusion Device.

As an interesting piece of data, the odorous total volatile organic sulfur compounds produced by the 5-pass cake during storage are plotted along with the total coliform density measured for the same cake in Figure 5-7. Similar to previous results, these data sets mimic each other. It should be noted that the cake stored for odor analysis was stored at 25°C, whereas the cake stored for regrowth was stored at 35°C, which may explain the slight offset in the peaks of the graphs.

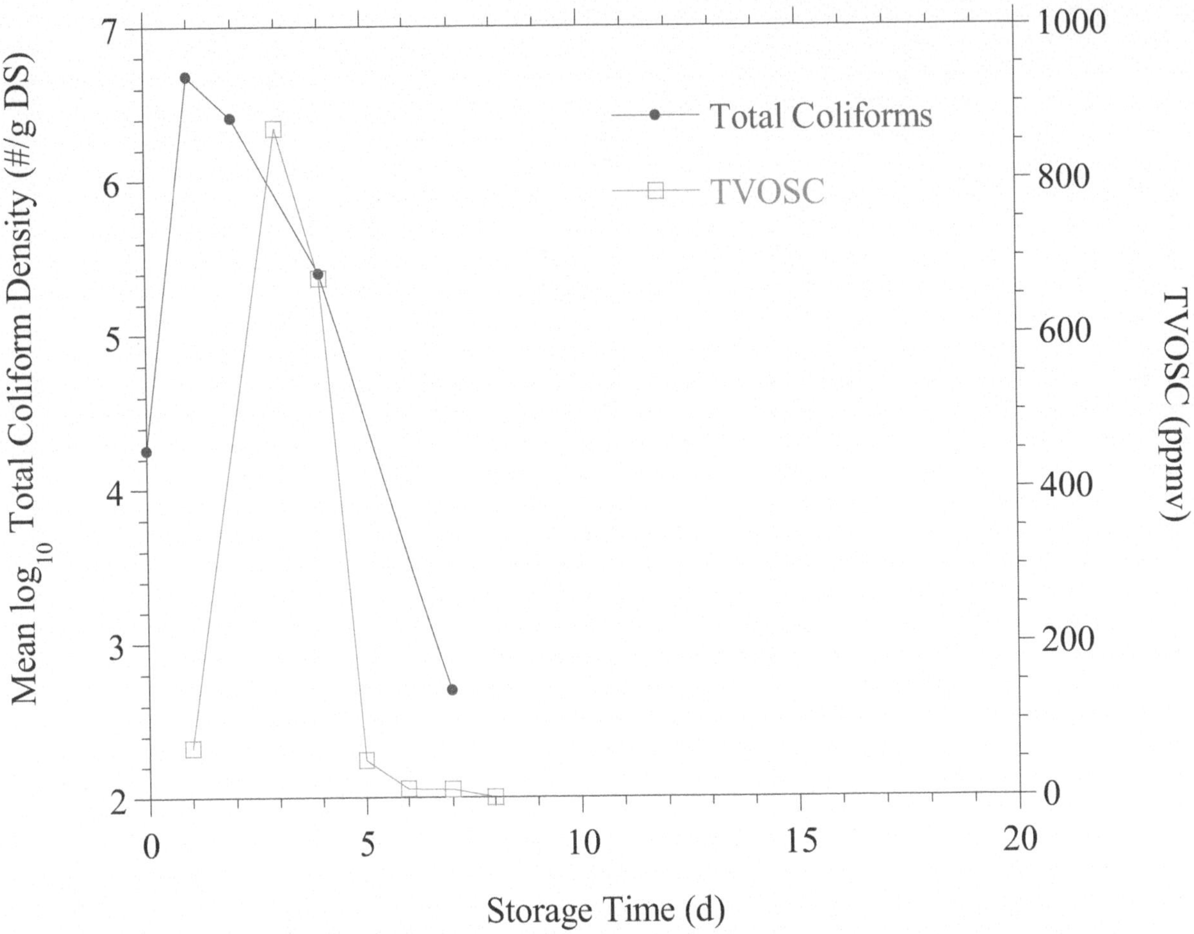

Figure 5-7.Total Coliforms Growth and Decay and Production of Odorous TVOSCs During Cake Storage for the High Solids Centrifuge Simulation Produced Cake with 5 Passes Through System.

CHAPTER 6.0

DO PATHOGENIC BACTERIA DENSITIES INCREASE AFTER DEWATERING?

6.1 Introduction

A number of enteric viral, bacterial and other pathogens can be present in wastewaters, and one important goal of wastewater treatment is to reduce the potential pathogenic activity of the wastewater. Currently, wastewater treatment plants rely on "indicator organisms" to determine the treatment process efficiency in terms of pathogen removal. The "indicator organisms" that wastewater treatment plants most often employ are fecal coliforms (FC) and *E. coli*. These organisms are present in the intestines of warm-blooded animals and in municipal wastewaters. As shown in the previous chapters, significant increases in indicator densities can be measured after dewatering of anaerobically digested biosolids compared to densities just prior to dewatering. The research has suggested that these increases are due to the presence of non-culturable bacteria prior to dewatering that are "reactivated" and then "regrow" after dewatering. The non-culturable and SI phenomenon has been described in the literature extensively for bacteria, especially related to drinking water, the food industry, as well as water quality. SI of viruses and protozoa is not expected in biosolids, therefore, the focus of the work is on bacteria.

A critical research need related to SI and regrowth is to determine if this phenomenon is also occurring with pathogenic bacteria of interest or is it only occurring for just indicator bacteria. The objectives of this research are to determine if potentially pathogenic bacteria are increasing in anaerobically digested biosolids after dewatering. The results from this work will provide the needed information to develop the path forward for the subsequent phase of the research. For example, if the bacterial pathogens are found to reactivate and regrow, the research could investigate mitigation of the problem for both Class A and Class B facilities. If they do not reactivate and regrow, then the field would be better served with research that examines the appropriate indicators for biosolids and regulatory requirements.

An additional benefit of the research is that it will provide six more sampling events in which the SI of indicator bacteria have been investigated since both FC and *E. coli* will be measured.

6.2 Additional Indicators and Potential Pathogenic Bacteria in Biosolids

There are a number of different bacterial pathogens and/or indicators that may warrant exploration based on the recommendations of the December, 2003 WERF Workshop on Indicators for Pathogens in Wastewater, Biosolids, and Stormwater and in the National Research Council's 2002 publication, *Biosolids Applied to Land: Advancing Standards.* The U.S. EPA has also provided additional insight into bacteria that the project should potentially include for testing (Meckes, personal communication). A summary of the potential bacteria to examine is provided in Table 6-1. The rationale for including or excluding each of these bacteria in the testing is provided below. Interestingly, a number of these organisms have been reported to enter a reversible non-culturable state. As a result, they may be subject to SI and/or regrowth after dewatering, similar the indicator bacteria fecal coliform and *E. coli*.

Table 6-1. Potential Bacterial Indicators and Pathogens that May be Present in Biosolids Along with the Type of Analytical Result that Can be Obtained.

Indicators and Pathogens	Reported to become reversibly non-culturable	Results Type
Fecal coliforms	Yes	Quantitative
E. coli	Yes	Quantitative
E. coli 0157:H7	Yes	Presence/Absence
Enterococci	Yes	Quantitative
Salmonella	Yes	Quantitative
Shigella	Yes	Presence/Absence
Aeromonas	Yes	Quantitative
Aerobic spore formers	No	Quantitative
Listeria monocytogenes	Yes	Presence/Absence
Campylobacter	Yes	Presence/Absence
Legionella spp	Yes	Quantitative
Helicobacter pylori	Yes	Quantitative

Fecal Coliform (FC). This group of organisms is currently used as one of the standards for Class A and Class B biosolids. In addition, they have been shown to reactivate and regrow after dewatering. Therefore, collecting FC results at the same time as measuring other pathogens will provide some valuable information on their use as an indicator organism.

***E. coli* (EC).** This is one of the most widely used fecal indicator organism in the water and wastewater industry worldwide. Research has shown that significant densities of potentially non-culturable *E. coli* can be present after digestion, and these non-culturable bacteria can be reactivated during some dewatering processes and regrow during cake storage. Similar to FC, collecting *E. coli* data will provide some valuable information on their use as an indicator organism.

***Escherichia coli* 0157:H7.** *E. coli* 0157:H7 is a pathogenic member of the fecal coliform group of organisms and will be included in this study. This organism has been responsible for disease outbreaks from both human and animal wastes (Rice, 1999), and the infective dose is thought to be very low, possibly fewer than 10 organisms (FDA, 2006). It occurs in domestic wastewater and has been detected in biosolids. In addition, researchers have reported that *E. coli* 0157 can enter a reversible, non-culturable state (Wang and Doyle, 1998; Mizunoe et al., 1999; Reissbrodt et al., 2002) and may therefore be reactivated and regrow after dewatering similar to the indicator bacteria.

Enterococci. This group of organisms is found in the intestinal tract of all warm-blooded animals, which includes humans. They are hardier than fecal coliforms and have been considered as an alternative indicator to fecal coliforms in wastewater and a good candidate for evaluating SI/regrowth in biosolids. This is currently the indicator organism used in the evaluating the microbiological quality of recreational waters. In addition, they are the leading cause of nosocomial infections, and several species are problematic due to their antibiotic resistance (Moellering, 1992; He and Jiang, 2005). *Enterococcus faecalis* has been shown to become reversibly non-culturable, and therefore may be able to reactivate and regrow after dewatering similar to the indicator bacteria FC and EC (Heim et al., 1002; Pfeffer et al., 2006; Lleo et al., 1998; and Lleo et al., 2001).

Salmonella spp. *Salmonella* is one of the principal pathogens of concern in biosolids and was included in the U.S. EPA 503 regulations. It is present in raw sewage and present in treated biosolids, although generally at low densities. It is used as one of the organisms for regulatory requirements related to Class A and Class B biosolids. *Salmonella* exposure is typically through contaminated foods such as raw meats, poultry and eggs, and can cause gastrointestinal distress including nausea, vomiting, diarrhea, cramps, and fever, and its infective dose can be as low as 15-20 bacteria (FDA, 2006). *Salmonella* have been shown to become non-culturable with the ability to be resuscitated (Reissbrodt et al., 2002; Gupte et al., 2003).

Shigella spp. This organism has a very low infective dose in humans (as few as 10 cells) and is considered as one of the principal pathogens of concern in domestic wastewaters and biosolids and was considered in the establishment of the U.S. EPA 503 regulations. Although there are concerns with the methodology for enumerating *Shigella* spp., providing presence/absence information will be valuable for the wastewater industry (Meckes, personal communication). *Shigella* have been reported to become non-culturable (see review by Oliver, 2005).

Aeromonas spp. The organism *Aeromonas* has been listed as an emerging bacterial pathogen. However, there is little data available on it occurrence in biosolids. *Aeromonas* is associated with warm blooded animals and is being considered by the U.S. EPA as a possible indicator of fecal contamination, although it is not considered a fecal bacterium (U.S. EPA, 2006). *Aeromonas* can cause diarrheal illness as well as non-intestinal infection, such as wound infections, septicemia, and endocarditis. HML, Inc. completed a holding-time study for the U.S. EPA and was able to isolate very high numbers of *Aeromonas* from wastewater. Several researchers have reported that *Aeromonas hydrophilia* could enter a reversible viable but non-culturable state (Wai et al., 2000; Sun et al., 2000), although Rahman et al (2001) reported results that showed the non-culturable state was not reversible. Since, the *Aeromonas* spp. can potentially become non-culturable, they represent a good target organism to examine for SI and regrowth in biosolids. In addition, preliminary testing of the methodology for enumeration of *Aeromonas* appears to provide reproducible results in biosolids.

Spore-formers: Aerobic endospores are ubiquitous in biosolids with densities reported of approximately 10^5 CFU/g dry weight. Meckes and Rhodes (2004), recently found bacterial aerobic endospores and *Clostridium perfringens* endospores have similar densities in alkaline treated biosolids. Also, due to the ease of analysis, bacterial aerobic endospores may be preferable to *Clostridium perfringens* endospores for biosolids screening. Including the spore formers may provide some valuable information on understanding the reasons for higher densities being measured after dewatering because their densities would not be expected to change appreciably through the digestion process.

Listeria monocytogenes. *Listeria sp.* have been detected in raw sewage and shows variable resistance to biosolids treatment processes (De Luca et al., 1998). Studies have found that *Listeria sp.* was not reduced by biological oxidation and in some cases become concentrated in biosolids (Geuenich and Muller, 1984). Anaerobic digestion does, however, reduce the numbers of *Listeria* (De Luca et al., 1998). Testing for *Listeria* was performed using a presence/absence test on two samples.

Campylobacter. *Campylobacter jejuni* is a gram-negative bacterium that can cause gastroentitis, and is typically transmitted by food, such as undercooked chicken or raw milk. The infective dose is estimated to be around 400-500 organisms, and the FDA considers *C. jejuni* to be a leading cause of bacterial diarrhea in the U.S., although, *C. jejuni* is relatively easily inactivated by heat, high oxygen concentration, and drying. *C. jejuni* is very difficult to culture, isolate, and quantify in the laboratory due to its fastidious nature, and therefore quantitative results will not be determined for this organism. However, two samples will be tested for presence/absence if a suitable method can be found. Researchers have reported that *C. jejuni* can enter a non-culturable state, suggesting that this sampling may show a similar sudden increase or SI of these organisms.

Legionella. *Legionella* is an organism that is ubiquitous in the environment. This organism has been detected at high concentrations in biosolids and in bioaerosols at wastewater treatment plants. Three cases of Legionnaires Disease in the U.S. have been associated with potting soil (MMWR, 2000). *Legionella* grows in hot water, such as cooling towers or hot water distribution systems and we can, therefore, assume that it may grow in warm, self composting organic masses. The growth of the organism could increase organism density to levels that represent a potential risk from aerosol transmission. Due to the difficulty and expense for enumerating this organism, it will not be included in the testing for the project.

Helicobacter pylori. *H. pylori* has been found to be associated with gastric and duodenal ulcers. The digestive tract of humans is a primary reservoir of *H. pylori* and therefore sewage pollution is a possible route of infection. It has emerged as an important risk factor for gastric cancer. There is no data about the fate of *H. pylori* in biosolids. However, due to limited recovery with the current methods, at best, only qualitative results may become available for this organism. Due to this uncertainty, this organism will not be analyzed as part of this project.

6.3 Reseach Approach

6.3.1 Sampling of Utilities

This phase of the project collected samples from five different utilities (with one site being sampled twice), and four different locations at each utility. Samples were taken in triplicate to provide enough data for statistical analysis. An additional Class B facility was sampled but the digester feed sample was not collected. All the facilities utilized anaerobic digestion. The locations for sampling included digester influent, digester effluent/feed to dewatering, dewatered cake, and a cake sample stored for 24 hours. These sampling locations are shown in Figure 6-1. Sampling of the post-centrifugation sample will be delayed after the pre-centrifugation sample to account for the solids retention time in the centrifuge. This was done to attempt to collect similar comparative samples.

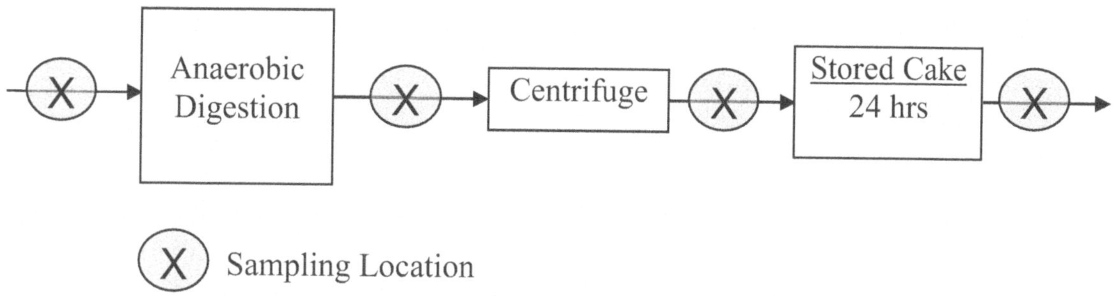

Figure 6-1. Sampling Locations for Bacterial Testing at Utilities With Anaerobic Digestion.

Samples were collected using asceptic techniques by the same person (Dr. Yen-Chih Chen) throughout the entire project. This was done to ensure consistency of sampling, and that the correct techniques were used for collecting, processing, and preparing the samples for shipping. Requiring the same person for sampling is a critical quality control measure for reliable and defendable data. Samples were generally collected late in the day to minimize shipping and holding times. After collection, the samples were immediately placed on ice and shipped overnight delivery to HML for analysis of the different bacteria by culturing techniques. In addition, a split set of samples were shipped to Bucknell University for molecular testing. Once at HML, a split sample of the cake solids was taken and stored for 24 hours at room temperature, and analyzed after this storage period for potential regrowth.

6.3.2 Sample Analysis – Culturing Methods

All bacterial analyses using culturing methods were performed at HML. Hoosier Microbiological Laboratory (HML) is a woman-owned business enterprise (GS24FOOO8M), a well-recognized NELAP accredited laboratory that is involved in all aspects of microbiological testing and research. Dr. Donald Hendrickson is a recognized expert in bacteriology and is a core team participant. Since little data are available for several of these bacteria in biosolids, initial testing of two facilities was used to determine the bacteria to analyze for the following sampling events. For the first two facilities, the following nine bacteria were analyzed:

1. Fecal coliform
2. *E. coli*
3. *E. coli* 0157
4. *Enterococci faecalis*
5. *Salmonella*
6. *Shigella*
7. *Listeria*
8. *Aeromonas*
9. Aerobic endospores
10. *Campylobacter*

Based on the results from this testing, a target list of bacteria was chosen for testing at the remaining facilities. The bacteria that were chosen was determined by the project team as well as from input from U.S. EPA, WERF, and the PSC. Therefore, the actual bacteria sampled for all the plants was determined based on some preliminary results.

6.3.3 Sample Analysis – Molecular Methods

In addition to the culturing method, *E. coli* and *Enterococci* were enumerated using molecular methods, and a pathogen screen was performed to analyze for the presence or absence of DNA copies of *Campylobacter jejuni*, *Shigella* spp., *Salmonella* spp., *E. coli* O157:H7, *Aeromonas* spp., and *Listeria* spp. according to the method of Fukushima et al., (2003). The *E. coli* enumeration was performed using real time polymerase chain reaction (RT-PCR) already developed at Bucknell University (Chen et al., 2006). *Enterococci* was enumerated by RT-PCR using an adaptation of the method of He and Jiang (2005). An MPN-PCR method for *Salmonella* enumeration was also used in two sample sets in which the *Salmonella* PCR method was used to confirm the presence of *Salmonella* in addition to the culturing method.

6.4 Results

6.4.1 Plant 1 – Pre-pasteurization Followed by Mesophilic Digestion

November, 2006 Sampling. The first plant was sampled three times over the course of the study, a period of about six months. This plant utilizes pre-pasteurization followed by mesophilic digestion and centrifuge dewatering as described in Chapter 3.0. The process information is provided in Table 3-3. The first sample was performed to trial some sampling and analytical methods, and collect some preliminary data. The bacterial density results for the preliminary trials are shown in Table 6-2. Similar to previous events, increases in FC and *E. coli* after dewatering was observed. Interestingly, *Enterococci* was also observed to increase immediately after dewatering. The *Salmonella* densities were very low in the raw sludge, about 16 MPN/4 g DS, and no detectable densities were measured after pre-pasteurization and digestion. Also, no increases or even detectable densities of *Salmonella* were measured after dewatering, indicating that *Salmonella* did not increase after dewatering. In addition to these bacterial densities, tests were conducted to quantify enteric virus and Helminth Ova on these samples. In all the samples from each location, the enteric viruses were all less than 1 pfu/4 g DS, and the Helminth ova were all less than 1 ovum/4 g DS.

Table 6-2. Results from Preliminary Testing on Additional Bacteria Species at Pre-Pasteurization Facility, November 2006.

SAMPLE Location	Fecal Coliform (MPN/g DS)	E. coli (MPN/g DS)	Enterococci (MPN/g DS)	Aeromonas (MPN/g DS)	Salmonella (MPN/4 g DS)
Pre Past Feed	39,177,727	7,479,113	904,954	5,743	15.2
Pre Past Feed	11,210,090	11,210,090	3,917,772	1,397	17.6
Pre Past Feed	218,530,833	16,658,261	3,093,642	969	16
Digester Feed	<4.5	<4.5	5.0	<0.45	<2
Digester Feed	<4.4	<4.4	91	1.0	<2
Digester Feed	<4.4	<4.4	14.3	0.5	<2
Centrifuge Feed	<11.3	<11.3	253.9	338.9	<5
Centrifuge Feed	<9.0	<9.0	39.4	172.7	<4
Centrifuge Feed	<8.6	<8.6	9.6	165.6	<3.8
Cake	8,072	8,072	12,605	2.2	<0.29
Cake	7,908	7,908	24,871	3.7	<0.37
Cake	12,783	12,783	754	0.9	<0.29

January 2007 Sampling. A second sampling was performed approximately two months later, and in this sampling event a larger number of bacteria were tested for the full regime of analyses. The results from the enumeration of the bacteria are shown in Table 6-3. Similar to previous testing, the FC and *E. coli* showed very low densities after pre-pasteurization and digestion, followed by a large increase immediately after dewatering (about four orders of magnitude in this case), and another two orders of magnitude increase during storage for 24 hours.

Table 6-3. Summary of FC, *E. coli*, and *Enterococcus* Densities Enumerated by Culturing and RT-PCR Methods for Pre-Pasteurization Facility in January, 2007.

SAMPLE	FC (MPN/g DS)	*E. coli* (MPN/g DS)	*E. coli* (copies/g DS)	*Enterococcus* (MPN/g)	*Enterococcus* (copies/g DS)
Digester Feed A	10,283,781	7,493,468	122,140,616	5,386,937	159,118,418
Digester Feed B	23,306,470	23,306,470	106,798,125	10,225,852	224,683,597
Digester Feed C	33,951,312	33,951,312	112,191,969	10,864,968	208,681,857
Digester Effluent A	12.5	12.5	358,972	81.2	6,018,569
Digester Effluent B	<12.0	<12.0	1,143,499	180.5	3,526,030
Digester Effluent C	<11.3	<11.3	1,186,923	1,036	4,241,079
Centrifuge Cake A	34,246	34,246	879,876	6,529	4,631,902
Centrifuge Cake B	13,071	13,071	1,173,169	60,507	10,204,453
Centrifuge Cake C	>60,606	>60,606	1,299,998	2,329	7,836,977
24 hr Cake A	2,990,264	2,990,264		6,073,445	
24 hr Cake B	2,690,950	2,690,950		3,530,111	
24 hr Cake C	20,461,950	20,461,950		>6,037,735	

As shown in Table 6-4, the *Salmonella* remained below 1 MPN/4 g DS after pre-pasteurization and digestion, and no increased densities were measured after centrifugation or after 24 hours of cake storage. The *Aeromonas* spp. did not increase immediately after centrifugation, but they did increase by about two orders of magnitude during cake storage. The aerobic endospores increased by about an order of magnitude after dewatering with no substantial increases after 24 hours of storage.

Table 6-4. Summary of Salmonella, Aeromonas, and Aerobic Endospores Enumeration Results at Different Sample Locations for Pre-Pasteurization Facility Sampled January, 2007.

SAMPLE	*Salmonella* (MPN/4g DS)	*Aeromonas* (MPN/g DS)	*Enterococcus* (MPN/g DS)	Aerobic Endospores (cfu/g DS)
Digester Feed A	0.8	6	5,386,937	43,000
Digester Feed B	<0.8	3.6	10,225,852	65,000
Digester Feed C	<0.8	10.9	10,864,968	57,000
Digester Effluent A	<1.6	92.4	81.2	11,000
Digester Effluent B	<1.7	34.7	180.5	12,000
Digester Effluent C	<1.6	45.1	1,036	18,000
Centrifuge Cake A	<1.0	0.8	6,529	350,000
Centrifuge Cake B	<1.0	1.2	60,507	380,000
Centrifuge Cake C	<1.0	0.3	2,329	19,000
24h Storage Cake A	<1.0	3,463	6,073,445	170,000
24h Storage Cake B	<1.0	310	3,530,111	28,000
24h Storage Cake C	<1.0	>603	>6,037,735	250,000

The samples were also evaluated using presence/absence tests for *Listeria*, *Campylobacter*, *E. coli* 0157:H7, and *Shigella*. The results from these tests are shown in Table 6-5. *Listeria* was detected as "present" in two of the three cake samples, although not after 24 hours of storage. Similarly, *Campylobacter* was present prior to digestion, but not found to be present at any point after digestion. *E. coli* O157 was found to be present, but then was not confirmed with secondary culturing methods, suggesting if it was present, it could not be cultured by the MacConkey-Sorbitol procedure or that it was not actually present. Two of the three digester effluent samples were positive for *Shigella*, but none of the cake samples were positive.

Table 6-5. Summary of Presence/Absence Tests for Bacteria at Pre-Pasteurization Facility, Sampled January, 2007.

SAMPLE	*Listeria*	*Campylobacter*	*E.coli* O157	*Shigella*
Digester Feed A	N.D.	P	NC	N.D.
Digester Feed B	N.D.	P	NC	N.D.
Digester Feed C	N.D.	P	NC	N.D.
Digester Effluent A	N.D.	N.D.	NC	P**
Digester Effluent B	N.D.	N.D.	NC	N.D.
Digester Effluent C	N.D.	N.D.	NC	P***
Centrifuge Cake A	N.D.	N.D.	NC	N.D.
Centrifuge Cake B	P	N.D.	NC	N.D.
Centrifuge Cake C	P	N.D.	NC	N.D.
24h Storage Cake A	N.D.	N.D.	NC	N.D.
24h Storage Cake B	N.D.	N.D.	NC	N.D.
24h Storage Cake C	N.D.	N.D.	NC	N.D.

N.D.: None detected; P: Present; NC: Not confirmed

* E.coli O157 did not confirm using MacConkey with Sorbitol and BIOLOG;

** Present for *Shigella flexneri, Shigella boydii,* and *Shigella sonnei*

*** Present for *Shigella sonnei*

The screen for bacterial DNA copies using RT-PCR showed that none of the bacteria except *Aeromonas* showed positive based on DNA copies as shown in Table 6-6. It should be noted that the detection limit for this method is relatively high, estimated in the 10^4 cells/g DS, therefore, a relatively high density of these organisms would need to be present to be detected.

Table 6-6. Summary of Presence/Absence Tests Using RT-PCR Screen for Bacteria at Pre-Pasteurization Facility Sampled January 2007.

SAMPLE	*Campylobacter jejuni*	*Shigella spp.*	*Salmonella spp.*	*E. coli* EPEC O157: H7	*Aeromonas spp.*	*Listeria spp.*
Digester Feed A	-	-	-	-	+	-
Digester Feed B	-	-	-	-	+	-
Digester Feed C	-	-	-	-	+	-
Digester Effluent A	-	-	-	-	+	-
Digester Effluent B	-	-	-	-	+	-
Digester Effluent C	-	-	-	-	+	-
Centrifuge Cake A	-	-	-	-	+	-
Centrifuge Cake B	-	-	-	-	+	-
Centrifuge Cake C	-	-	-	-	+	-

* PCR positive (+) or negative (-) tests were used for *Campylobacter jejuni, Shigella spp., Salmonella spp., E. coli* EPEC *O157:H7, Aeromonas spp., and Listeria, spp.*

April 2007 Sampling. The third sampling of the pre-pasteurization facility took place in April of 2007. For this sampling, several of the bacterial presence/absence tests had been dropped, such that only *E. coli* O157 and *Shigella* were tested for presence/absence, and *Clostridium* spp. was added to the enumeration procedures. The results for the FC, *E. coli*, and *Enterococcus* enumeration by both MPN and RT-PCR procedures are shown in Table 6-7. The FC and *E. coli* results were typical of previous samplings showing large increases immediately after dewatering followed by additional increases in densities after storage. A very similar pattern was found with *Enterococcus* as shown in Table 6-7.

Table 6-7. Summary of FC, *E. coli*, and *Enterococcus* Densities Enumerated by Culturing and RT-PCR Methods for Pre-Pasteurization Facility in April, 2007.

SAMPLE	FC (MPN/g DS)	*E. coli* (MPN/g DS)	*E. coli* (copies/g DS)	*Enterococcus* (MPN/g DS)	*Enterococcus* (copies/g DS)
Digester Feed A	3,671,304	232,739	72,878,286	2,361,826	186,585,559
Digester Feed B	5,382,583	4,408,458	71,651,567	358,666	139,437,595
Digester Feed C	22,172,285	5,396,265	64,570,050	665,244	211,664,263
Digester Effluent A	15	15	<279,997	105	9,319,214
Digester Effluent B	<16.4	<16.4	<279,997	62	16,178,021
Digester Effluent C	<15	<15	<279,997	57	20,121,413
Centrifuge Cake A	6,223	3,920	<290,144	1,629	5,488,590
Centrifuge Cake B	1,171	1,171	<290,144	7,871	13,354,886
Centrifuge Cake C	19,364	6,157	<290,144	6,157	10,433,870
24h Storage Cake A	56,871,487	19,160,484	NT	982,674	NT
24h Storage Cake B	8,719,690	8,719,690	NT	179,360	NT
24h Storage Cake C	32,432,137	4,591,272	NT	383,901	NT

The results from enumeration using culturing methods for the other bacteria (*Salmonella*, *Aeromonas*, Aerobic Endospores, and *Clostridium*) are shown in Table 6-8. The *Salmonella* densities in the digester feed were greater than previous sampling, however, the densities measured after digestion, dewatering, and storage of the cake all were below the detection limit. The *Aeromonas* densities did not change appreciable after digestion, but they did decrease after dewatering and did not increase during storage. The aerobic endospores decreased slightly during digestion, and then increased by about an order of magnitude after dewatering, and slightly increased during storage of the cake. A similar patter was found for *Clostridium*. The results from the presence/absence test for *E. coli* O157 and *Shigella* are summarized in Table 6-9 and show that these bacteria are not present after digestion and are not reactivated or regrown after dewatering and storage, respectively. These results are consistent with the sampling three months prior. Finally, the RT-PCR screen also did not show any of the bacteria to be present except *Aeromonas* as shown in Table 6-10.

Table 6-8. Summary of *Salmonella*, *Aeromonas*, and Aerobic Endospores Enumeration Results at Different Sample Locations for Pre-Pasteurization Facility Sampled April, 2007.

SAMPLE	*Salmonella* (MPN/4g DS)	*Aeromonas* (MPN/g DS)	Aerobic Endospores (cfu/g DS)	*Clostridium* (cfu/g DS)
Digester Feed A	139.9	53.5	86,000	136,000
Digester Feed B	29	84	100,000	124,000
Digester Feed C	34.7	107	96,000	109,000
Digester Effluent A	<2.0	267.5	17,000	7,000
Digester Effluent B	<2.4	98.7	15,000	9,300
Digester Effluent C	<2.2	108.3	19,000	7,800
Centrifuge Cake A	<0.9	1.4	135,000	149,000
Centrifuge Cake B	<0.9	1.7	119,000	120,000
Centrifuge Cake C	<0.9	2.3	123,000	101,000
24h Storage Cake A	<0.9	3.8	200,000	590,000
24h Storage Cake B	<0.9	3.5	163,000	310,000
24h Storage Cake C	<0.9	25	280,000	270,000

Table 6-9. Summary of Presence/Absence Tests for Bacteria at Pre-Pasteurization Facility, Sampled April, 2007.

SAMPLE	*E.coli* O157	*Shigella*
Digester Feed A	NC	N.D.
Digester Feed B	NC	P**
Digester Feed C	NC	N.D.
Digester Effluent A	NC	N.D.
Digester Effluent B	NC	N.D.
Digester Effluent C	NC	N.D.
Centrifuge Cake A	NC	N.D.
Centrifuge Cake B	NC	N.D.
Centrifuge Cake C	NC	N.D.
24h Storage Cake A	NC	N.D.
24h Storage Cake B	NC	N.D.
24h Storage Cake C	NC	N.D.

* E.coli O157 did not confirm using MacConkey with Sorbitol and BIOLOG; N.D.: None detected; P: Present
**Note--Present for Shigella sonnei 1& 2

Table 6-10. Summary of Presence/Absence Tests Using RT-PCR Screen for Bacteria at Pre-Pasteurization Facility Sampled April 2007.

SAMPLE	*Campylobacter jejuni*	*Shigella spp.*	*Salmonella spp.*	*E. coli* EPEC O157: H7	*Aeromonas spp.*	*Listeria spp.*
Digester Feed A	-	-	-	-	+	-
Digester Feed B	-	-	-	-	+	-
Digester Feed C	-	-	-	-	+	-
Digester Effluent A	-	-	-	-	+	-
Digester Effluent B	-	-	-	-	+	-
Digester Effluent C	-	-	-	-	+	-
Centrifuge Cake A	-	-	-	-	+	-
Centrifuge Cake B	-	-	-	-	+	-
Centrifuge Cake C	-	-	-	-	+	-

[*] PCR positive (+) or negative (-) tests were used for *Campylobacter jejuni, Shigella spp., Salmonella spp., E. coli* EPEC *O157:H7, Aeromonas spp., and Listeria, spp.*

6.4.2 Thermophilic Digestion with TPAD

This plant was not sampled previously as part of this phase of the research, but had been previously sampled during the first phase. The plant is a Class A plant that used time-temperature to meet Class A digestion requirements, followed by a temperature phased anaerobic digestion, with the final process being mesophilic digestion. The densities of FC, *E. coli*, and *Enterococus* measured by both culturing methods and RT-PCR are shown in Table 6-11 for each replicate sample. The results show a typical pattern of SI and regrowth of these indicator bacteria. The densities of the other bacteria measured by culturing methods only are shown in Table 6-12. The densities of *Salmonella* were low entering the digestion process, and remained below detectable levels after digestion and dewatering. *Aeromonas* densities decreased during digestion by about one order of magnitude, decreased further after dewatering, but then increased after 24 hours of cake storage. In contrast the density of aerobic endospores did not change appreciably during digestion, but increased by an order of magnitude following dewatering. The results from the presence/absence tests for the culturing based methods are summarized in Table 6-13. The results indicate the *Shigella* and *Campylobacter* were not present in the samples, and that *E. coli* O157 could not be confirmed in the samples. Several of the samples showed presence of *Listeria*, but not for all the replicates. The centrifuge cake sampled was one out of the three, and after storage two out of the three showed presence of *Listeria*. The RT-PCR screen for *Campylobacter jejuni*, *Shigella* spp., *Salmonella* spp., *E. coli* O157: H7, *Aeromonas* spp., *Listeria* spp. did not show any to be present (within the detection limit) except for *Aeromonas* which tested positive in the digester feed and digester effluent but not the cake sample (data not shown).

Table 6-11. Summary of FC, *E. coli*, and *Enterococcus* Densities Enumerated by Culturing and RT-PCR Methods for TPAD Facility in February, 2007.

SAMPLE	FC (MPN/g DS)	*E. coli* (MPN/g DS)	*E. coli* (copies/g DS)	*Enterococcus* (MPN/g DS)	*Enterococcus* (copies/g DS)
Digester Feed A	292,678	292,678	7,545,527	141,750	38,389,608
Digester Feed B	281,607	281,607	2,005,686	1,175,285	3,379,955
Digester Feed C	554,987,275	554,987,275	9,046,333	470,724	56,858,348
Digester Effluent A	206	206	<80,000	205.5	598,684
Digester Effluent B	565	146	<80,000	104.3	1,424,986
Digester Effluent C	2,055	900	<80,000	54.1	757,542
Centrifuge Cake A	1,024	1,024	<80,000	25,696	1,423,080
Centrifuge Cake B	1,378	317	<80,000	9,162	1,032,784
Centrifuge Cake C	32,388	32,388	<80,000	11,101	<80000
24h Storage Cake A	75,208,556	75,208,556	NT	1,120,518	NT
24h Storage Cake B	>80,808,080	>80,808,080	NT	121,106	NT
24h Storage Cake C	6,187,285	6,187,285	NT	44,342	NT

Table 6-12. Summary of *Salmonella*, *Aeromonas*, and Aerobic Endospores Enumeration Results at Different Sample Locations for TPAD Facility Sampled February, 2007.

SAMPLE	*Salmonella* (MPN/4g DS)	*Aeromonas* (MPN/g DS)	Aerobic Endospores (cfu/g DS)
Digester Feed A	1	2,879	50,000
Digester Feed B	3.1	404.6	100,000
Digester Feed C	<0.9	144.6	9,200
Digester Effluent A	<1.1	17.1	39,000
Digester Effluent B	<1.1	20.2	70,000
Digester Effluent C	<1.1	13.3	31,000
Centrifuge Cake A	<1.2	0.8	350,000
Centrifuge Cake B	<1.4	1.7	220,000
Centrifuge Cake C	<1.2	1	560,000
24h Storage Cake A	<1.2	324	260,000
24h Storage Cake B	<1.3	61.7	1,090,000
24h Storage Cake C	<1.2	58.2	470,000

Table 6-13. Summary of Presence/Absence Tests for Bacteria at TPAD Facility, Sampled February, 2007.

SAMPLE	*Listeria*	*Campylobacter*	*E.coli* O157	*Shigella*
Digester Feed A	N.D.	N.D.	NC	N.D.
Digester Feed B	N.D.	N.D.	NC	N.D.
Digester Feed C	P	N.D.	NC	N.D.
Digester Effluent A	N.D.	N.D.	NC	N.D.
Digester Effluent B	N.D.	N.D.	NC	N.D.
Digester Effluent C	N.D.	N.D.	NC	N.D.
Centrifuge Cake A	P	N.D.	NC	N.D.
Centrifuge Cake B	N.D.	N.D.	NC	N.D.
Centrifuge Cake C	N.D.	N.D.	NC	N.D.
24h Storage Cake A	N.D.	N.D.	NC	N.D.
24h Storage Cake B	P	N.D.	N.D.	N.D.
24h Storage Cake C	P	N.D.	NC	N.D.

* E.coli O157 did not confirm using MacConkey with Sorbitol and BIOLOG; N.D.: None detected; P: Present

6.4.3 Thermophilic Digestion with Reactors in Series (Thermophilic Plant 2)

The thermophilic digestion facility with reactors in series was also sampled for bacterial pathogens and additional indicators. This facility was the same process described in Chapter 3.0 and given the descriptor "Thermophilic Plant 2", and the process information is given in Table 3-2. The densities of the indicator bacteria measured in samples at the different locations by both culturing and RT-PCR are shown in Table 6-14. Similar to previous sampling, the FC and *E. coli* densities were low after digestion, and no SI or regrowth was observed. The *Enterococcus* results were similar to *E. coli* for the culturing based methods, although the RT-PCR results showed high densities of DNA copies, suggesting SI potential existed for *Enterococcus.* The densities of *Salmonella, Aeromonas, Clostridium,* and aerobic endospores measured by culturing methods are summarized in Table 6-15. The densities of *Salmonella* were found to be greater than 153 MPN/4 g DS in the digester feed, but were below detection after digestion, and no SI or regrowth was observed for *Salmonella* after dewatering and storage, respectively. *Aeromonas* densities were low after digestion, and decreased below detection after dewatering and remained below the detection limit in the cake after storage. *Clostridium* and aerobic endospores showed a similar pattern, in that after digestion, their densities were in the range of 10^4 cfu/g DS, and both increased by about an order of magnitude after dewatering, did not change appreciably during cake storage. Presence/absence tests were performed for *E. coli* O157 and *Shigella*. No *Shigella* were detected in any of the samples, and although the samples showed *E. coli* O157 as being present, they could not be confirmed using the MacConkey-Sorbitol procedure, data not shown. Similar to previous samples, the RT-PCR screen for *Campylobacter jejuni*, *Shigella* spp., *Salmonella* spp., *E. coli* O157: H7, *Aeromonas* spp., *Listeria* spp. did not show any to be present (within the detection limit of the test) except for *Aeromonas* which tested positive in the digester feed and digester effluent but not the cake sample (data not shown).

Table 6-14. Summary of FC, *E. coli*, and *Enterococcus* Densities Enumerated by Culturing and RT-PCR Methods for Thermophilic Reactors in Series Facility in February, 2007.

SAMPLE	FC (MPN/g DS)	*E. coli* (MPN/g DS)	E. coli (copies/g DS)	*Enterococcus* (MPN/g DS)	*Enterococcus* (copies/g DS)
Digester Feed A	25,867,666	25,867,666	140,220,312	797,904	234,491,050
Digester Feed B	25,867,666	25,867,666	127,345,961	4,104,333	600,944,193
Digester Feed C	41,043,333	41,043,333	118,198,396	797,904	460,766,231
Digester Effluent A	<10.5	<10.5	627,467	<10.5	5,545,823
Digester Effluent B	<10.5	<10.5	96,120	<10.5	12,075,318
Digester Effluent C	<10.5	<10.5	387,714	<10.5	15,809,981
Centrifuge Cake A	<6.3	<6.3	<80,000	<6.3	8,678,125
Centrifuge Cake B	<6.4	<6.4	<80,000	<6.4	2,360,284
Centrifuge Cake C	<6.4	<6.4	<80,000	<6.4	11,766,782
24h Storage Cake A	<6,369	<6,369	NT	<6,369	NT
24h Storage Cake B	<6,431	<6,431	NT	<6,431	NT
24h Storage Cake C	<6,515	<6,515	NT	<6,515	NT

Table 6-15. Summary of *Salmonella*, *Aeromonas*, and Aerobic Endospores Enumeration Results at Different Sample Locations for Thermophilic Reactors in Series Facility Sampled February, 2007.

SAMPLE	*Salmonella* (MPN/4g DS DS)	*Aeromonas* (MPN/g DS)	Aerobic Endospores (cfu/g DS)	*Clostridium* (cfu/g DS)
Digester Feed A	>153	1,198	46,000	101,000
Digester Feed B	>153	671	45,000	104,000
Digester Feed C	>153	397	80,000	160,000
Digester Effluent A	<1.4	4.2	26,000	22,000
Digester Effluent B	<1.4	1	33,000	23,000
Digester Effluent C	<1.4	2.4	45,000	21,000
Centrifuge Cake A	<0.8	<0.1	300,000	270,000
Centrifuge Cake B	<0.8	<0.1	300,000	200,000
Centrifuge Cake C	<0.8	<0.1	290,000	230,000
24h Storage Cake A	<0.8	<0.6	310,000	270,000
24h Storage Cake B	<0.8	<0.6	300,000	280,000
24h Storage Cake C	<0.8	<0.6	330,000	290,000

6.4.4 Thermophilic Digestion with Single Stage Reactors (Thermophilic Plant 1)

The single stage thermophilic process described in Chapter 3.0 as "Thermophilic Plant 1" was also sampled as part of the pathogen testing. A description of the process parameters for this plant can be found in Table 3-1. Although a thermophilic process is used, the plant is a Class B facility. The densities of the indicator bacterial are shown in Table 6-16 for both the culturing methods and the RT-PCR. The results showed that SI did not occur, but regrowth was significant after 24 hours of storage for the indicator bacteria. The densities of *Salmonella*, *Aeromonas*, *Clostridium*, and aerobic endospores measured by culturing methods are summarized in Table 6-17, and the results show that *Salmonella* remained below detection and *Aeromonas* remained low in all the samples. Both *Clostridium* and aerobic endospores increased by about an order of magnitude after centrifugation and increased slightly during storage. The results from the presence/absence testing showed that *Shigella* was not present in any of the samples, while *E. coli* O157 was detected as present, it could not be confirmed in any of the samples, as shown in Table 6-18. Similar to the previous testing, the pathogen screen using RT-PCR showed that *Campylobacter jejuni*, *Shigella* spp., *Salmonella* spp., *E. coli* O157: H7, *Aeromonas* spp., *Listeria* spp. did not show any to be present (within the detection limit of the test) except for *Aeromonas* which tested positive in the digester feed and digester effluent as well as two out of the three cake sample (data not shown).

Table 6-16. Summary of FC, *E. coli*, and *Enterococcus* Densities Enumerated by Culturing and RT-PCR Methods for Thermophilic Plant 1 Facility in February, 2007.

SAMPLE	FC (MPN/g DS)	*E. coli* (MPN/g DS)	*E. coli* (copies/g DS)	*Enterococcus* (MPN/g DS)	*Enterococcus* (copies/g DS)
Digester Feed A	89,148,333	89,148,333	60,981,049	1,264,717	357,170,205
Digester Feed B	7,285,260	7,285,260	62,007,400	1,722,652	233,218,565
Digester Feed C	55,289,325	27,161,050	49,928,037	1,749,125	243,166,276
Digester Effluent A	26,746	13,261	1,498,422	6,095	1,827,574
Digester Effluent B	41,708	26,746	1,525,068	6,095	1,929,454
Digester Effluent C	31,609	31,609	1,128,519	4,484	2,562,678
Centrifuge Cake A	509	323	19,296	50.9	674,190
Centrifuge Cake B	661	661	320,240	20.9	790,429
Centrifuge Cake C	210	80	102,872	27.9	586,429
24h Storage Cake A	28,240,907	28,240,907	NT	5,304,061	NT
24h Storage Cake B	50,295,721	50,295,721	NT	6,911,165	NT
24h Storage Cake C	49,828,578	49,828,578	NT	10,764,040	NT

Table 6-17. Summary of *Salmonella*, *Aeromonas*, Aerobic Endospores, and *Clostridium* Enumeration Results at Different Sample Locations for Thermophilic Plant 1 Sampled February, 2007.

SAMPLE	*Salmonella* (MPN/4 g DS)	*Aeromonas* (MPN/g DS)	Aerobic Endospores (cfu/g DS)	*Clostridium* (cfu/g DS)
Digester Feed A	>1,650	6	38,000	36,000
Digester Feed B	>1,399	5.9	34,000	29,000
Digester Feed C	>1,609	102.9	42,000	94,000
Digester Effluent A	19.3	9	20,000	13,200
Digester Effluent B	19.3	1.4	19,000	9,700
Digester Effluent C	17.6	3.7	12,000	11,600
Centrifuge Cake A	<0.8	0.7	210,000	140,000
Centrifuge Cake B	<0.8	0.5	250,000	142,000
Centrifuge Cake C	<0.8	1.2	220,000	177,000
24h Storage Cake A	<0.8	7.9	820,000	210,000
24h Storage Cake B	<0.8	<0.6	580,000	190,000
24h Storage Cake C	<0.8	6.5	700,000	220,000

Table 6-18. Summary of Presence/Absence Tests for Bacteria at Thermophilic Plant 1.

SAMPLE	*E.coli* O157	*Shigella*
Digester Feed A	NC	N.D.
Digester Feed B	NC	N.D.
Digester Feed C	NC	N.D.
Digester Effluent A	NC	N.D.
Digester Effluent B	NC	N.D.
Digester Effluent C	NC	N.D.
Centrifuge Cake A	NC	N.D.
Centrifuge Cake B	NC	N.D.
Centrifuge Cake C	NC	N.D.
24h Storage Cake A	NC	N.D.
24h Storage Cake B	NC	N.D.
24h Storage Cake C	N.D.	N.D.

* *E.coli* O157 did not confirm using MacConkey with Sorbitol and BIOLOG; N.D.: None detected; P: Present

6.4.5 Mesophilic Digestion (Mesophilic Plant 3)

A facility with conventional, single stage mesophilic digestion (Mesophilic Plant 3 in Chapter 3.0) was also sampled for additional bacterial indicators and pathogens. The densities of indicator bacteria measured by culturing methods and in the case of *E. coli* and *Enterococcus* also measured by RT-PCR are shown in Table 6-19. The FC and *E. coli* did not increase immediately after centrifugation, but increased substantially during storage, suggestive of little SI, but significant regrowth. The *Enterococcus* densities increased after centrifugation by less than an order of magnitude, and also increased further during storage, but not as much as the *E. coli* densities. The densities of *Salmonella*, *Aeromonas*, *Clostridium*, and aerobic endospores measured by culturing methods are summarized in Table 6-20. The *Salmonella* densities entering digestion were greater than the method could be measured based on the number of dilutions used, so the value estimate is unknown, but greater than approximately 130 MPN/4 g DS. Unlike the previously sampled thermophilic processes, measurable densities of *Salmonella* were enumerated after digestion, with an average of about 35.5 MPN/4 g of DS. After dewatering, the *Salmonella* densities averaged about 44 MPN/4 g of DS, a slight increase. Further increases were measured after storage, and the densities increased above the detection threshold for the number of dilutions specified for the test (U.S. EPA Method 1682, June 2005). As a result, the *Salmonella* density in the cake is only known to be greater than 210 MPN/4 g DS. The aerobic endospores and *Clostridium* showed a similar pattern to other samples, with an order of magnitude increase after dewatering, followed by additional increases during storage. The presence/absence tests for *E. coli* O157 and *Shigella* were similar to previous tests. *Shigella* was not found to be present in any of the samples, while *E. coli* O157 was present but not confirmed in the samples. The RT-PCR assay also did not show the different bacteria to be definitively present, except for *Aeromonas*, which was present in all the samples.

Table 6-19. Summary of FC, *E. coli*, and *Enterococcus* Densities Enumerated by Culturing and RT-PCR Methods for Mesophilic Plant 3.

SAMPLE	FC (MPN/g DS)	*E. coli* (MPN/g DS)	*E. coli* (copies/g DS)	*Enterococcus* (MPN/g DS)	*Enterococcus* (copies/g DS)
Digester Feed A	5,277,500	3,346,468	66,477,286	661,281	206,864,811
Digester Feed B	5,170,218	3,298,562	58,053,272	247,625	256,646,540
Digester Feed C	3,937,363	1,494,666	60,912,100	99,727	266,214,085
Digester Effluent A	543,200	164,550	2,207,152	80,450	72,530,411
Digester Effluent B	1,545,000	780,888	2,352,032	>88,888	56,321,992
Digester Effluent C	683,842	259,578	2,220,734	84,684	36,514,217
Centrifuge Cake A	416,451	253,980	310,539	146,217	16,846,151
Centrifuge Cake B	338,882	250,765	678,435	221,408	16,493,824
Centrifuge Cake C	156,088	104,139	299,068	156,088	28,612,238
24h Storage Cake A	5,500,983	5,500,983		703,895	
24h Storage Cake B	7,836,323	7,836,323		208,049	
24h Storage Cake C	14,860,061	14,860,061		423,429	

Table 6-20. Summary of *Salmonella, Aeromonas*, Aerobic Endospores, and *Clostridium* Enumeration Results at Different Sample Locations for Mesophilic Plant 3.

SAMPLE	*Salmonella* (MPN/4 g DS)	*Aeromonas* (MPN/g DS)	Aerobic Endospores (cfu/g DS)	*Clostridium* (cfu/g DS)
Digester Feed A	>201	146	47,000	37,000
Digester Feed B	>201	135.1	41,000	42,000
Digester Feed C	>128.6	106.9	35,000	50,000
Digester Effluent A	38.1	8.2	38,000	18,000
Digester Effluent B	39.6	42.3	44,000	22,000
Digester Effluent C	28.7	9.1	57,000	21,000
Centrifuge Cake A	45	6	390,000	206,000
Centrifuge Cake B	44	6.1	250,000	200,000
Centrifuge Cake C	44	8.7	340,000	280,000
24h Storage Cake A	>210	9.3	1,370,000	430,000
24h Storage Cake B	>210	5.4	1,270,000	430,000
24h Storage Cake C	>210	6	1,380,000	370,000

Table 6-21. Summary of Presence/Absence Tests Using RT-PCR Screen for Bacteria at Mesophilic Plant 3.

SAMPLE	*Campylobacter jejuni*	*Shigella spp.*	*Salmonella spp.*	*E. coli* EPEC O157: H7	*Aeromonas spp.*	*Listeria spp.*
Digester Feed A	-	-	-	-	+	-
Digester Feed B	-	-	-	-	+	-
Digester Feed C	-	-	-	-	+	-
Digester Effluent A	-	-	-	-	+	-
Digester Effluent B	-	-	-	-	+	-
Digester Effluent C	-	-	-	-	+	-
Centrifuge Cake A	-	-	-	-	+	-
Centrifuge Cake B	-	-	-	-	+	-
Centrifuge Cake C	-	-	-	-	+	-

[*] PCR positive (+) or negative (-) tests were used for *Campylobacter jejuni, Shigella spp., Salmonella spp., E. coli* EPEC *O157:H7, Aeromonas spp., and Listeria, spp.*

An additional experiment was performed in which the cake was stored for an extended period of seven days, and the analysis was performed using the U.S. EPA Method 1682 and also an MPN-PCR method for comparison. The MPN-PCR method was used to reduce the time for confirming *Salmonella* from the MSRV "halo-positive" colonies. The MPN-PCR confirmation requires only a few additional hours beyond the MSRV test, while the U.S. EPA Method 1682 requires an additional three days of culturing to confirm *Salmonella.* The two methods showed 100% agreement in the confirmation of *Salmonella*, which verified, at least for research purposes, that the MPN-PCR method could be used to confirm *Salmonella*. The regrowth of

Salmonella during storage is shown in Figure 6-2. Unfortunately, the method limit is around 200 MPN/4 g DS due to the dilutions prescribed by the method are mainly for Class A biosolids, and low densities are expected. However, the results do show that regrowth of *Salmonella* does occur, and the densities remained greater than 193 MPN/4 g DS during the seven days of storage.

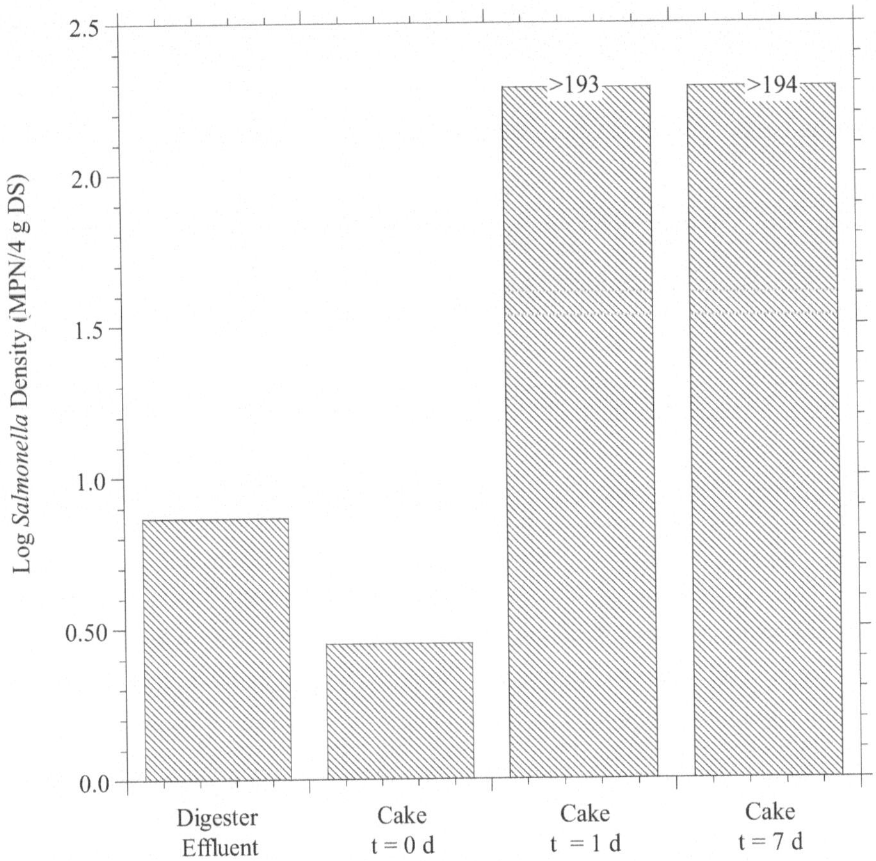

Figure 6-2. *Salmonella* Measured Before and During Storage of Centrifuged Cake from Mesophilic Plant 3.

6.4.6 Mesophilic Digestion (Mesophilic Plant 6)

An additional sampling was performed from a conventional mesophilic plant, and in this case, the testing focused on FC, *E. coli*, and *Salmonella* such that a more comprehensive regrowth data set could be collected. In addition, the site was performing a trial with side-by-side centrifuge and belt filter press dewatering, which provided a valuable opportunity for comparison. The sampling also focused on post-digestion since the number of samples that could be processed was limited, and for *Salmonella* analysis, the MPN-PCR method was used for enumeration as described previously. In addition, an extended set of dilutions was performed for the *Salmonella* testing, since the dilutions given in the method had a maximum limit of about 200 MPN/g DS, and it was found previously that greater densities were possible in Class B cakes (see Figure 6-2). Figure 6-3 shows the regrowth curves for *E. coli* and *Salmonella* during storage of the cake at 35°C. The pattern of *E. coli* regrowth and die-off during storage of the centrifuge cake was fairly typical of previous tests. Interestingly, the belt filter press cake also had significant regrowth during cake storage, much higher than any previous belt filter press cake that had been sampled. The cake solids from the belt filter press was 25.4% and the centrifuge cake was 24.1%, which may explain, in part, the differences. The *Salmonella* densities also showed a regrowth and a slow die-off during the week of storage. At time zero, the dilution set was not great enough to obtain estimates of the *Salmonella* densities, but based on the data, the belt filter press and centrifuge cake samples had densities greater than 3600 and 3800 MPN/4 g DS, respectively. These values were used for plotting purposes. Similarly, the belt filter press cake after 1 d of storage had a *Salmonella* density greater than 36,400 MPN/4 g DS, and this value was used for plotting purposes.

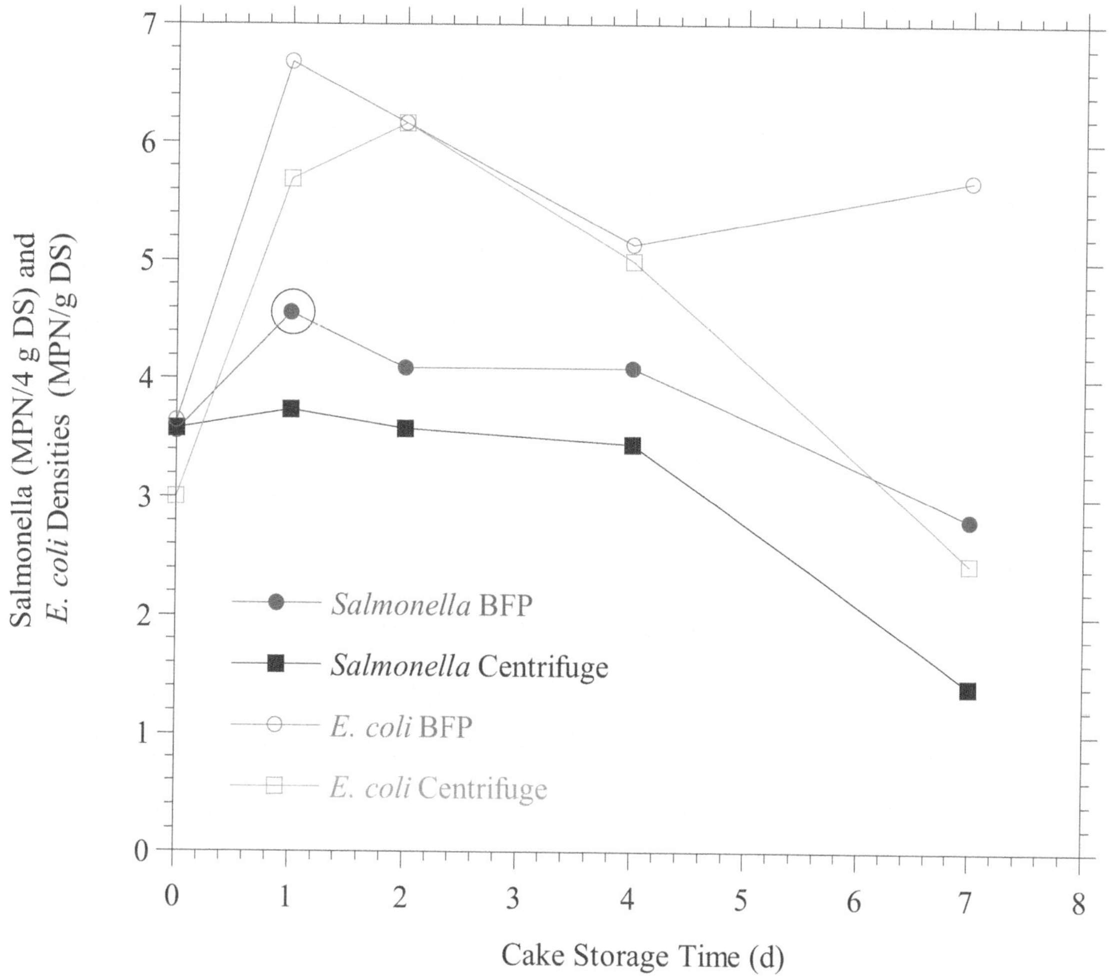

Figure 6-3. Regrowth Curves for *E. coli* and *Salmonella* in Cakes from Side-by-Side Centrifuge and Belt Filter Press Dewatering Equipment at Mesophilic Plant 6 (Note: Circled Values Indicate Actual Density Was Greater Than Plotted Value).

6.5 Discussion

6.5.1 Pathogens in Undigested Sludges

The results from this testing provide some interesting insights into indicators and potentially pathogenic bacteria as they related to digestion and biosolids. However, the difficulty in assessing the results arises in that few standard and verified methods exist for enumerating most of these different bacterial indicators and pathogens in biosolids. The only tests that are approved by the U.S. EPA are for fecal coliforms and *Salmonella*. A summary of the digester influent data for *Salmonella, E. coli, Enteroccus, Aeromonas*, and *Shigella* is shown in Table 6-22. The data show that the mean *E. coli* density is in the range of 10^6 to 10^8 per gram dry solids for all samples. Similarly, the mean *Enterococcus* densities are the range of 10^5 to 10^6 and the *Aeromonas* range from 7 to 2700 MPN/g DS. The mean *Salmonella* densities ranged from less than one to greater than 1600 MPN/4 g DS, and unfortunately, for many of the samples, the *Salmonella* densities exceeded the detection threshold of the test which has a dilution set designed more for Class A biosolids in which the *Salmonella* density is expected to be low. No correlation was found between any of the indicators. The results also provide a comparison of influent changes for three samples collected at the same plant (Pre-Pasteurization) over a five month period. The mean densities of *Salmonella, E. coli* and *Enterococcus* were within about an order of magnitude, while *Aeromonas* ranged over three orders of magnitude.

Table 6-22. Summary of Digester Influent Results from Each Plant.

	***Salmonella* (MPN/4 g DS)**	***E. coli* (MPN/g DS)**	***Enterococcus* (MPN/g DS)**	***Aeromonas* (MPN/g DS)**	***Shigella***
	15.2	7,479,113	904,954	5,743	NT
Pre-Past - Nov	17.6	11,210,090	3,917,772	1,397	NT
	16	16,658,261	3,093,642	969	NT
	0.8	7,493,468	5,386,937	6	N.D.
Pre-Past - Jan	<0.8	23,306,470	10,225,852	3.6	N.D.
	<0.8	33,951,312	10,864,968	10.9	N.D.
	139.9	232,739	2,361,826	53.5	N.D.
Pre-Past - Apr.	29	4,408,458	358,666	84	P**
	34.7	5,396,265	665,244	107	N.D.
	1	292,678	141,750	2,879	N.D.
TPAD	3.1	281,607	1,175,285	404.6	N.D.
	<0.9	554,987,275	470,724	144.6	N.D.
	>153	25,867,666	797,904	1,198	N.D.
Thermo-2	>153	25,867,666	4,104,333	671	N.D.
	>153	41,043,333	797,904	397	N.D.
	>1,650	89,148,333	1,264,717	6	N.D.
Thermo-1	>1,399	7,285,260	1,722,652	5.9	N.D.
	>1,609	27,161,050	1,749,125	102.9	N.D.
	>201	3,346,468	661,281	146	N.D.
Meso-3	>201	3,298,562	247,625	135.1	N.D.
	>128.6	1,494,666	99,727	106.9	N.D.

The digester influent samples only had one "presence" for *Shigella* out of the 21 samples collected, and this sample was confirmed to be present for *Shigella sonnei*. None of the analyses for *E. coli* O157, although showing presence, could be confirmed with additional culturing methods. Overall, the results seem to indicate that the bacterial pathogens in the undigested sludges are relatively low compared to the indicator bacteria such as *E. coli*, and *Enterococcus*.

6.5.2 SI/Regrowth of Pathogens After Digestion and Dewatering

The results indicate that for Class A processes, there does not appear to be any significant SI and regrowth of pathogens. For example, placing an emphasis on *Salmonella* because it is a bacterial pathogen that has an U.S. EPA-approved method for its enumeration in biosolids, none of the Class A processes had detectable *Salmonella* after digestion, and the *Salmonella* density remained below detection after dewatering and after 24 hours of cake storage, indicating no SI and regrowth. Although, two of these plants had SI and/or regrowth of the indicators *E. coli* and *Enterococcus.* These results agree with Iranpour and Cox (2006) who reported regrowth of fecal coliforms but not *Salmonella* after dewatering of thermophilically digested biosolids.

For the Class B plants sampled (Thermo-1, Mesophilic-3, and Mesophilic-6), regrowth of *Salmonella* occurred in the two mesophilically digested biosolids, but not the thermophilically digested biosolids. Extended storage time testing at the Mesophilic-6 facility also showed that regrowth of *Salmonella* occurred, and densities exceeding 10^4 MPN/4 g DS were measured during storage, and remained near this level during the week of storage. The results did not indicate that SI was occurring at these facilities, but rather regrowth of *Salmonella* was responsible for the increases.

Overall, the results are supportive of Class A biosolids having low or no bacterial pathogens, even though indicator bacteria may increase after dewatering, this did not occur for the bacterial pathogens measured in this study. The results provide support for the U.S. EPA time-temperature regulations to meet Class A requirements for *Salmonella* even though it may not be met for indicators such as fecal coliforms or *E. coli*. The data suggests that the FC may be a conservative indicator for pathogen destruction.

In the case of Class B biosolids, the results show regrowth of *Salmonella* does occur, similar to regrowth of the indicator bacteria. However, it is expected that Class B biosolids will have pathogens present, and the public health protection is through site restrictions associated with land applications, rather than a specific limit on pathogens in the biosolids. In other words, the restricted public access and exposure to the biosolids for land application of Class B would provide the public safety, and not a limit on the microbial densities in the biosolids.

The results can help guide future research related to this issue.

CHAPTER 7.0

PERSPECTIVES ON SI: MECHANISMS AND MITIGATION

7.1 Introduction

The results from this study and the previous WERF study on the subject have provided some initial information to help in understanding the issue of SI and regrowth. There has been significant discussion about the possible mechanisms related to why the sudden increase is occurring or being measured, and what are the possibilities for mitigating SI and regrowth. Section 7.2 reviews the different theories that have been promulgated to explain the sudden increase with an evaluation of both the WERF data and the data from the literature, and an analysis of the limitations of the data. Section 7.3 discusses the possibilities for mitigation of SI and regrowth.

7.2 Evaluation of Hypotheses for the Sudden Increase

As discussed in Chapter 1.0, several theories have been promulgated to explain the sudden increase. These include:

1. Statistical variability in the sampling;
2. False positives associated with enumeration techniques;
3. Contamination from the centrifuge;
4. Regrowth of bacteria after dewatering;
5. Improved culturability after dewatering, possibly due to:
   a. Floc breakup due to shear
   b. Removal of growth inhibitors during dewatering;
   c. Protozoan grazing;
   d. Reactivation of non-culturable bacteria followed by regrowth

Some of these theories have been evaluated and discussed in the literature, and a review of the different theories within the context of the WERF data and the literature is provided in the following sections.

7.2.1 Statistical Variability in the Sampling

Others have also proposed that the SIs are due to statistical variability. In other words, since the variability with indicator bacteria enumeration can be fairly high, this might result in the measured increases of bacterial densities after dewatering. In several recent publications, as well as this WERF project, the data collection was performed in such a way that statistical analysis could be performed (Qi et al., 2007; Monteleone, 2004; Higgins et al., 2006). Based on the statistical analysis, the researchers were able to demonstrate that the increases were statistically valid in many cases using a 95% confidence value as the criteria for statistical validity. Therefore, it is very reasonable to conclude that the increases measured after dewatering are not simply due to variability in the data, especially for thermophilic processes. However, for mesophilic processes, the SI is sometimes zero or up to one order of magnitude, which in some cases could be within the statistically variability of the test.

7.2.2 False Positives Results due to Enumeration Techniques

Qi et al. (2007) examined whether false positives could be resulting from Bacillus species that are present in the samples. Spiking experiments and other evaluations led them to conclude that this was not a likely cause of the sudden increase in after dewatering.

7.2.3 Contamination from the Centrifuge

Another theory that has been discussed is contamination and seeding from a centrifuge. In this theory, a thin layer of solids that exists between the bowl and scroll of a centrifuge is colonized by bacteria. Over time, this layer sheds bacteria into the bulk cake, increasing the density of the organisms measured after dewatering.

A data set reported by Hendrickson et al. (2001) appears to refute this theory. For example, Hendrickson et al. (2001) disinfected a centrifuge that was producing cakes with a significant sudden increase in indicator organism using a sodium hypochlorite solution prior to operating it with the digested solids. Immediately after disinfection, samples of digested cake were taken downstream of the centrifuge to be analyzed for fecal coliform. Fecal SI was observed in all the samples. This test seems to indicate that seeding of indicators does not appear be occurring from contamination from centrifuges. However, this could be occurring in some situations, and additional experimentation is needed to further investigate this possibility.

An experiment was performed as part of this research to examine the timing of the increase after centrifugation. The objective was to determine if the SI was truly "immediate" or if it required several hours. For example, most field samples were stored on ice for several hours prior to analysis by SCMs. Although on ice, it is conceivable that this time period was required for the SI to occur. In this test, the samples were collected from the centrifuge of the pre-pasteurization plant and analyzed/placed in media as quickly as possible. The samples were then analyzed every few hours after collection. The time required for sample collection and processing prior to putting the sample into the media was estimated as about 40 minutes, and therefore cannot be considered truly "immediate" but this was the shortest time interval that could be achieved. The *E. coli* densities measured prior to dewatering and then after dewatering at the various time intervals are shown in Figure 7-1. Interestingly, the increase occurred within the 40 minutes or so after dewatering, and the regrowth appears to not occur until after a few hours of a lag phase. The results could support the centrifuge contamination theory, since the increase was measured immediately, however, the lag period is interesting in that if the bacteria were growing rapidly in the centrifuge, a lag period would not be expected.

Additional work is needed to further examine this concept to determine if centrifuge contamination is contributing to the SI after dewatering.

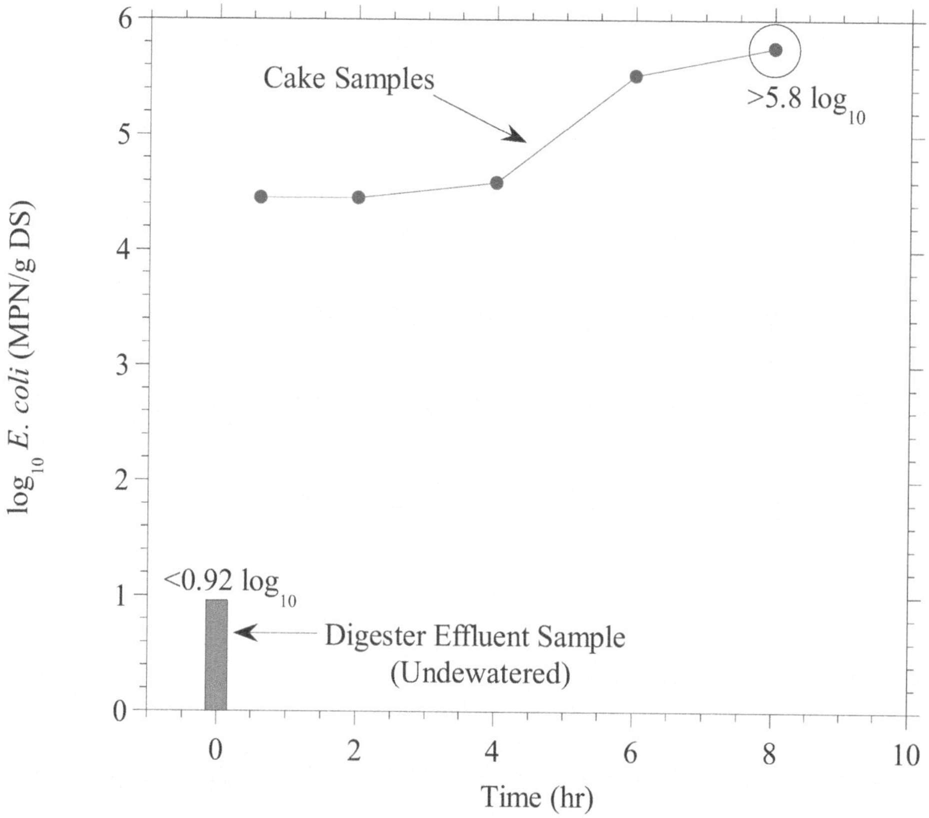

Figure 7-1. *E. coli* Densities Measured Immediately Before and After Dewatering with No Storage on Ice. Samples Obtained from the Pre-Pasteurization Plant.

7.2.4 Regrowth of Bacteria After Dewatering

One theory also proposed as a possible mechanism is that the sudden increase was due to regrowth of the bacteria after dewatering. In some cases, regrowth may contribute; however, it cannot explain the orders of magnitude increases measured in some cases after dewatering. Under the most ideal conditions, *E. coli* can double about every 20 minutes. If it is assumed that for the purpose of analysis that the *E. coli* double every 20 minutes, regrowth would not explain the one or more order of magnitude increase after dewatering since the retention time in a centrifuge is typically on the order of 20 minutes. Also, results from this research have shown that the doubling times of *E. coli* in the cake after dewatering and the sudden increase is on the order of 2-4 hours, much slower than the ideal conditions that produce a doubling time of 20 minutes. Several other researchers also concluded that regrowth likely did not explain the large, sudden increase measured in some situations (Qi et al., 2007; Iranpour et al., 2003). Therefore, it is possible that regrowth contributes to some of the increase, but cannot explain the larger (>1 log) increases measured after dewatering at some plants.

7.2.5 Improved Culturability of Bacteria After Dewatering

Another explanation for the increased densities measured after dewatering could be due to improved culturability after dewatering. The improved or increased culturability could due to several mechanisms, and these different mechanisms are discussed in the following sections.

7.2.5.1 Floc Break-up or Disaggregation

The floc-breakup concept is that if large numbers of the target bacteria are embedded in a floc, this floc needs to be broken up to accurately enumerate all the organisms. If the floc is not broken up, then the enumeration technique might underestimate the total numbers of the organism. For example, taking an aliquot of a liquid sludge sample with one floc that had a high number of bacteria in it, and placing the aliquot into an MPN tube or plated would result in an enumeration equivalent to just one bacterium. This effect would potentially underestimate the total numbers compared to if the bacteria were dispersed throughout the medium prior to placing them into MPN tubes or onto plates for enumeration. To minimize this problem, Method 1680 promulgated by the U.S. EPA for enumeration of FC in biosolids (and the method used in this study) specifies that the samples should be homogenized in a sterile blender on high speed for one or two minutes prior to placing in the MPN tubes (U.S. EPA, 2005a). For example, liquid samples can be blended directly, while 30 grams of cake are placed in 270 mL of dilution buffer prior to blending (U.S. EPA, 2005a).

Despite the use of a blender for homogenizing the samples prior to enumeration, several researchers have hypothesized that the high shear associated with centrifuge dewatering could break up the flocs, and better disperse the bacteria, resulting in greater numbers that could be enumerated. If so, this would account for the greater culturable densities measured immediately after dewatering compared to just prior to dewatering. Results from the *Clostridium* and the aerobic spore formers discussed in Chapter 6 showed that the densities of these bacteria increased by about an order of magnitude after centrifuge dewatering. This is potential supporting evidence of the floc-breakup theory since it was believed that the densities of these bacteria should not change appreciably during the digestion and dewatering process due to the resistance of spores to environmental stressors. Although, it is also possible that the increased densities of spore formers was a result of the bacteria sporulating after dewatering.

In contrast, several researchers examined this effect of shear, and found that shearing of sludges in the laboratory did not increase the densities measured in sludges (Qi et al., 2007 and 2007; Cheung et al., 2003). Qi et al. (2007) sheared sludge using a kitchen blender for extended periods but did not measure an increase in FC densities after blending compared to a control with no blending. Similarly, they also examined the use of sonic dispersion and separately they tried a cake extrusion process, but neither increased the densities of the indicator bacteria measured in the sludges (Qi et al., 2007). Cheung et al. (2003) examined the floc dispersal theory by homogenizing sludge samples using a stomacher or with an ultrasonic treatment. Neither of these floc dispersal techniques increased the indicator bacterial densities measured in the liquid sludge samples. Interestingly, homogenizing cake samples with a stomacher increased the density measured by about 0.5 log units. This is contrary to the floc dispersal theory, although potentially consistent with the dewatering process. Although centrifuge dewatering can impart shear to the cake, polymer is added for the purpose of aggregating the flocs into larger particles such that they dewater well, and resist floc breakup during centrifugation. Therefore, the net effect of dewatering should be an overall aggregation of the floc, rather than a floc breakup and release of bacteria. The increase in densities after homogenizing the cake samples supports this concept. In addition, research in this study showed that for a thermophilic plant the densities were below detection limits after digestion, yet after dewatering, the densities were greater than 10^4 per gram DS. If the floc breakup theory were true, this result would not be expected. At least a few of the samples after digestion should show positive results rather than none.

To further investigate this hypothesis, several liquid and full-scale centrifuge cake samples were evaluated for particle size after homogenization by blending of the liquid or 30 grams of cake in 270 mL of dilution buffer. In one test, the particle size was measured at different time intervals to examine the impact of blending time. Results from Meso-3 liquid and dewatered cake samples are shown in Figure 7-2. As shown in the graph, the mean particle sizes (for triplicate analyses) of the liquid samples tend toward smaller particles compared to the centrifuged cake samples which have greater numbers of particles in the larger size ranges (>10 µm). This data then does not support the floc breakup theory, and supports Cheung et al.'s (2003) result that showed addition shear of the cake led to greater enumeration. The results suggest that inadequate homogenization of the cake may actually be resulting in underestimation of the bacteria in the cake samples rather than vice-versa.

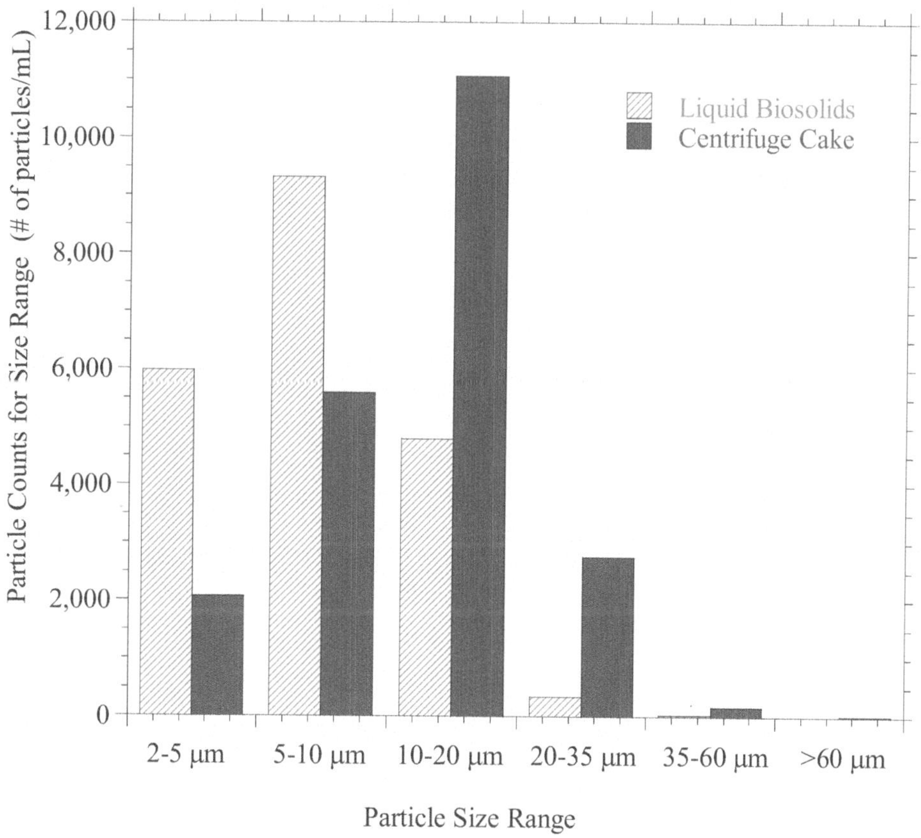

Figure 7-2. Particle Size Distribution of Liquid (Undewatered) and Centrifuge Dewatered Cake Samples After Blending of Meso-3 Samples. The Data Represent the Mean of Triplicate Analyses.

A similar set of particle size analyses were performed for the Pre-Pasteurization plant samples since this plant often had a fairly high sudden increases after dewatering. The results, shown in Figure 7-3, were very similar to those for Meso-3. After homogenization by blending, the liquid samples prior to centrifugation had a smaller particle size distribution compared to the centrifuge cake samples. The cake sample had greater numbers of the larger particle sizes (>10 µm). It should be noted that the solids concentration after blending were very similar, less than 10% difference, after taking into consideration the dilution of the cake in the buffer.

To further investigate the effect of blending, a trial was performed in which the particle sizes were measured at regular intervals during blending. The results from the liquid biosolids samples from the Pre-pasteurization facility are shown in Figure 7-4. Interestingly, the number of particles in the 10 µm and greater particle size ranges decreased during blending, while the particle counts in the 2-5 µm and the 5-10 µm size ranges increased. This shows that the larger particles are breaking up to form smaller particles. Examination of Figure 7-4 also suggests that the disaggregation is not yet complete after 2 minutes of blending since the counts in the 2-5 µm range are still increasing, while the counts in the 5-10 µm range had plateaued after 2 minutes of blending. Similar results were found for the cake samples as shown in Figure 7-5, except that the particle counts greater than 20 mm decreased during dewatering, while the 2-5, 5-10, and 10-20 µm particle counts increased during blending. Also, the particle size ranges of 2-5 and 5-10 µm were increasing steadily during blending and did not level off, suggesting that additional blending would results in a smaller particle size distribution, especially for the cake samples.

Although the data is not shown, a third plant (Meso-6) was also analyzed for particle size distribution after dewatering, and the results were very similar to Meso-3 and the Pre-pasteurization plant results. The undewatered biosolids had a particle size distribution with greater numbers of smaller particles compared to the centrifuged cake.

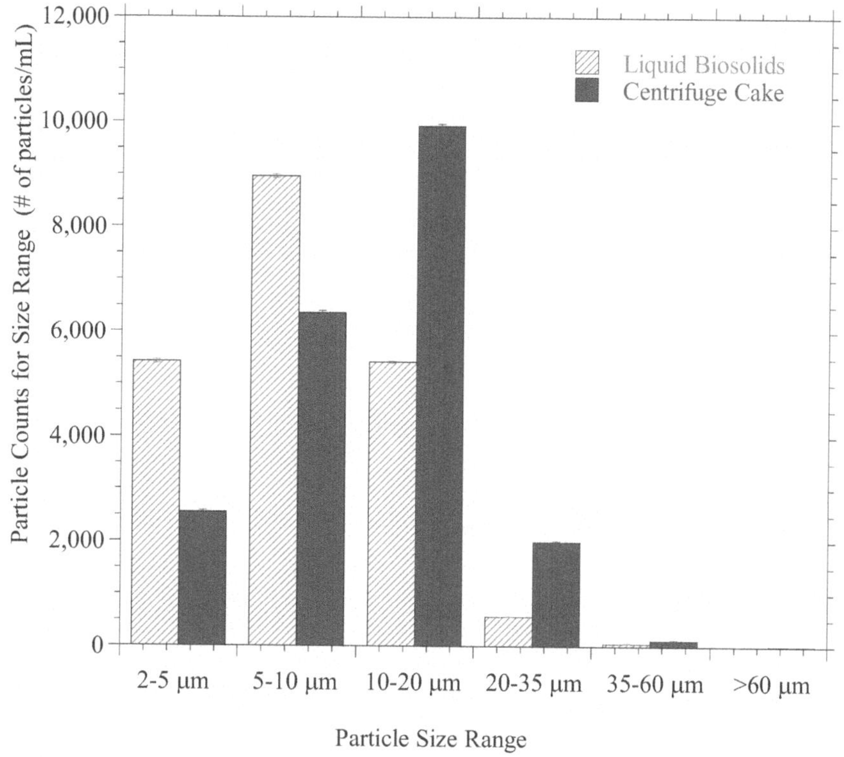

Figure 7-3. Particle Size Distribution of Liquid and Dewatered Cake Samples After Blending of Pre-Pasteurization Samples. The Data Represent the Mean of Triplicate Analyses and Error Bars Represent One Standard Deviation.

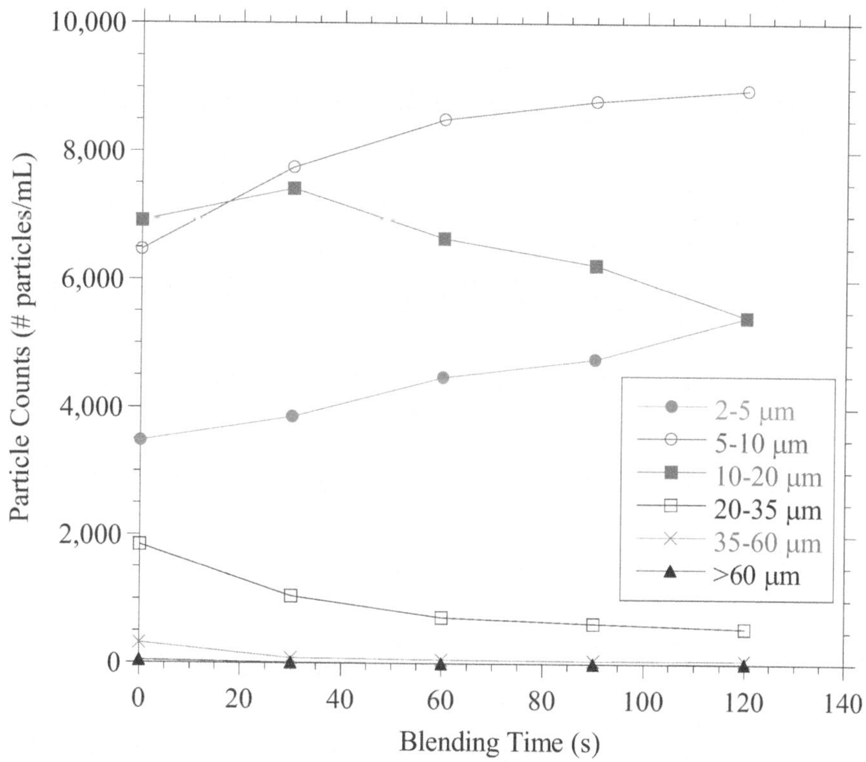

Figure 7-4. Changes in Particle Counts for Different Particle Sizes During Blending of Liquid (Undewatered) Biosolids from Pre-Pasteurization Plant.

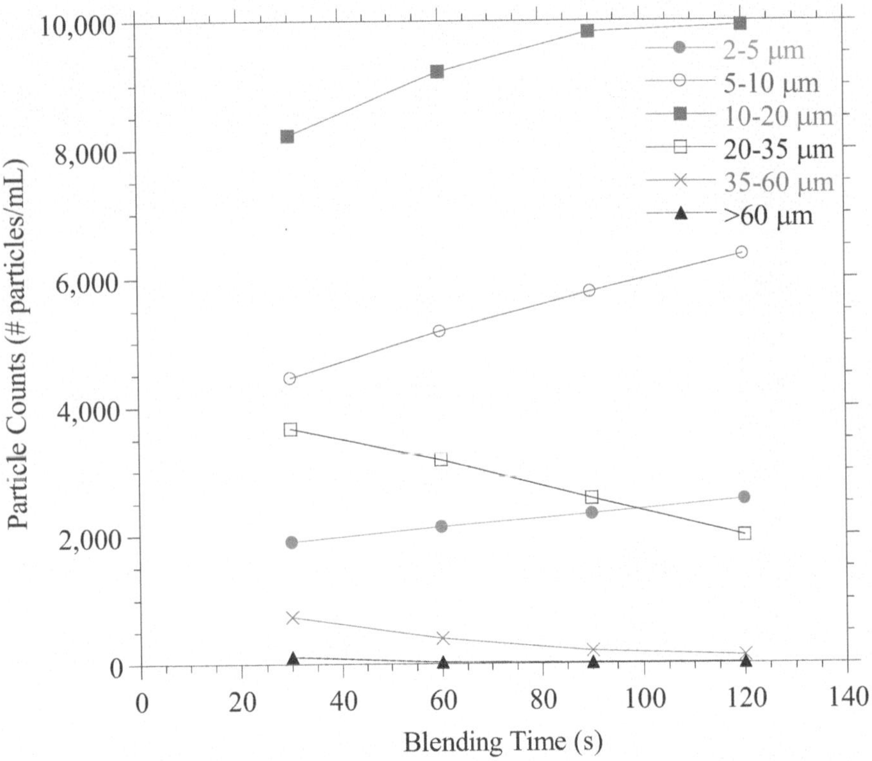

Figure 7-5. Changes in Particle Counts for Different Particle Sizes During Blending of Centrifuge Dewatered Cake Samples from Pre-Pasteurization Plant.

Finally, the qPCR results also do not support the floc dispersal theory. If floc dispersal were the reason for the increases, the qPCR results should also increase in a similar manner to the SCM results due to better sampling/dispersal of the flocs. This did not occur in any of the cases when qPCR was performed. Similarly, the floc dispersal theory should apply to all bacteria, and increases should be measured for the different species unless other effects were happening. In each case of the pathogen sampling events, all six sampling events found lower densities of *Aeromonas* after dewatering compared to prior to dewatering.

The results from our research agree with other researchers (Qi et al., 2007 and 2007, Cheung et al., 2003) that indicate floc dispersal is likely not the main mechanism for the increased densities after dewatering. In fact, the results appear to indicate that additional floc dispersal would be needed for better enumeration of bacteria in the cake samples. This is supported by the particle size data as well as the data from Cheung et al. (2003) that found additional dispersion of the cake resulted in greater densities being enumerated by culturing methods. Additional testing is needed to further investigate this mechanism and also optimize laboratory blending/shearing for accurate enumeration.

7.2.5.2 Removal of Growth Inhibitors During Dewatering

Another possible mechanism is that during centrifuge dewatering, a growth inhibitor may be removed. This may be a result of the addition of polymer and removal of the centrate, separating it from the solids. Experiments in this study that examined the effect of cell washing and lab centrifuging did not result in the large increases that would be expected if inhibitors were present that prevented bacterial growth on the media. Two other studies also reported that lab centrifugation did not result in an increase in the indicator densities (Cheung et al., 2003; Iranpour et al., 2003). In addition, one experiment was performed that added centrate back to the growth media, and this led to an increase in the numbers that were enumerated (Higgins et al., 2007). This would not be expected if inhibitors were preventing growth. For some utilities, this mechanism may be important, and additional testing may be warranted to further investigate this possibility.

7.2.5.3 Ingestion and Release of Bacteria by Protozoa

Another possible mechanism to explain the increase in densities measured after dewatering would be due to ingestion of the bacteria by protozoa during digestion, however, the bacteria are not consumed by the protozoa, and are then released after dewatering. Release could be due to breakup or shear of the protozoa. Several researchers have shown that some bacteria, such as *E. coli* and *Salmonella* can survive predation by protozoa (King et al., 1988; Brandl et al., 2005; Tezcan-Merdol et al., 2004), to be later released into the environment. It is possible then that bacteria could be ingested by protozoa during digestion, but the bacteria survive ingestion. While ingested, the bacteria would not be culturable by typical culturing methods. Centrifuge dewatering could then release the bacteria which are then cultured. This hypothesis was not tested during the study, and is in need of further investigation.

7.2.5.4 Reactivation of Non-culturable Bacteria Followed by Regrowth

The final theory to be evaluated is that the bacteria enter a reversibly non-culturable state during digestion, and are then reactivated by the dewatering process. The concept of reversibly non-culturable (RNC) has been around for several decades. The first demonstration of RNC was within the context of sub-lethal injury. Researchers demonstrated that bacteria, including indicator bacteria, which were sub-lethally injured would not grow on selective media used for their enumeration, but the bacteria could be coaxed to grow when the growth media was amended or growth conditions were adjusted (Lin et al., 1976; Martin et al., 1976; Green et al., 1977; Stuart et al., 1977; LeChavallier et al., 1982; McFeters et al., 1982; McDonald et al., 1983; McFeters et al., 1986). The sensitivity to selective media was thought to be in part associated with oxidants such as peroxide, deoxycholate, or more non-specific stress associated with selective media (Barry et al., 1956; Martin et al., 1976; McFeters et al., 1982). In the stressed condition of sub-lethal injury, the bacteria would not be able to grow on the media. By supplementing the media with scavengers to remove these components such as catalase or pyruvate, researchers could significantly increase the numbers of bacteria that could be enumerated (Martin et al., 1976; McDonald et al., 1983; Mizunoe et al., 1999). Presumably, these media enhancement removed enough of the stresses associated with the media to allow the injured bacteria to repair and grow on the media, and subsequently be enumerated.

Another theory that has also developed to explain RNC bacteria and has similar outcomes is the viable but non-culturable (VBNC) concept. In this theory, bacteria enter a survival mode in response to stresses such as temperature, low substrates and nutrients, and chlorine exposure. In this VBNC state, bacteria have been reported to change their physiology to survive the stress (Berlin et al., 1999; Signoretto et al., 2000; Heim et al., 2002). Interestingly similar to the sub-lethal injury concept, as a result of being in the VBNC state, the bacteria do not grow well on the selective media used for enumeration, but media enhancements or changes in growth conditions can lead to a 'resuscitation' of the bacteria (Whitesides and Oliver, 1997; Mizunoe et al., 1999; Reissbrodt et al., 2000; Haro-Kudo, 2000; Grey and Steck, 2001; Reissbrodt et al., 2002; Gupte et al., 2003). These enhancements have included both pyruvate and catalase, which have also been shown to allow sub-lethally injured bacteria to be enumerated. Also, researchers have used alternate enumeration or viability assays to enumerate or investigate the viability of the bacteria that do not rely on their culturability. These methods have included the live/dead stain, RNA evaluations, and PCR analysis. Researchers have shown that VBNC bacteria are capable in some instances of causing infection *in vivo* (Colwell et al., 1985; Colwell et al., 1996; Pommepuy et al., 1996; Bogosian et al., 1998; Cappelier et al., 1999; Chaveerach et al., 2003). In contrast to these results, several researchers have reported that VBNC bacteria are not capable of causing infection (Caro et al, 1999; Kolling and Matthews, 2001). Therefore, it is not clear whether the bacteria in a reversibly non-culturable state are capable of infection, and additional research would be needed to further investigate this.

From a review of the literature, there is a large amount of evidence that demonstrates bacteria, including *E. coli* and FC, can become reversibly non-culturable (RNC). Whether the RNC state is due to sub-lethal injury or VBNC is very difficult to discern, since many of the outcomes are identical, and there continues to be debate about these mechanism in the literature. What is clear though, is bacteria can become RNC. The water, drinking water, and wastewater regulators have recognized this phenomenon, and the U.S. EPA has even promulgated a change in methodology for enumerating these bacteria which includes a resuscitation step and the use of mTEC media (U.S. EPA, 2005 - Methods 1603 and 1103.1). The occurrence of RNC bacteria in biosolids, is therefore, very plausible, but little research on this phenomenon in biosolids has been performed, except in this WERF-funded research. The data collected in this study appear to support the RNC concept to explain the sudden increase in FC and *E. coli*. An alternative culturing independent method for enumerating *E. coli* was used to test the hypothesis in addition to SCMs. This method, qPCR, enumerates *E. coli* by the number of copies of *E. coli* DNA found in the sample. Although there are several limitations to this method which are described below, this approach does provide a method to test the hypothesis. For example, if the hypothesis is correct, several outcomes would be expected from the qPCR results which include:

1. the qPCR results should be greater than the SCM results after digestion when the bacteria are thought to be RNC; the difference between the two would be considered the ‘reactivation potential’;
2. the qPCR measured densities should not increase significantly in samples collected immediately after dewatering;
3. after dewatering, the SCM results and qPCR should be in agreement if reactivation occurs.

The qPCR and SCM results support each one of these outcomes. For example, a plot or the reactivation potential, and the actual amount of reactivation experienced after dewatering is plotted in Figure 6-1. This is the same plot shown in Figure 3-30, but for convenience to the reader, it is shown here again. The reactivation potential was taken as the difference between the qPCR and SCM results after digestion and prior to dewatering, and this is plotted on the x-axis. The actual amount of reactivation, which is the difference between the SCM density measured after dewatering and the SCM density measured prior to dewatering is plotted on the y-axis. As shown in Figure 7-6, the correlation between the reactivation potential and the actual amount of reactivation is very good. This would only be expected if the qPCR densities were a measure of potentially RNC bacteria. Based on the slope of the line, about 80% of the reactivation potential is realized on average after dewatering. This may indicate that a fraction of the qPCR results represent non-viable or irreversibly non-culturable bacteria.

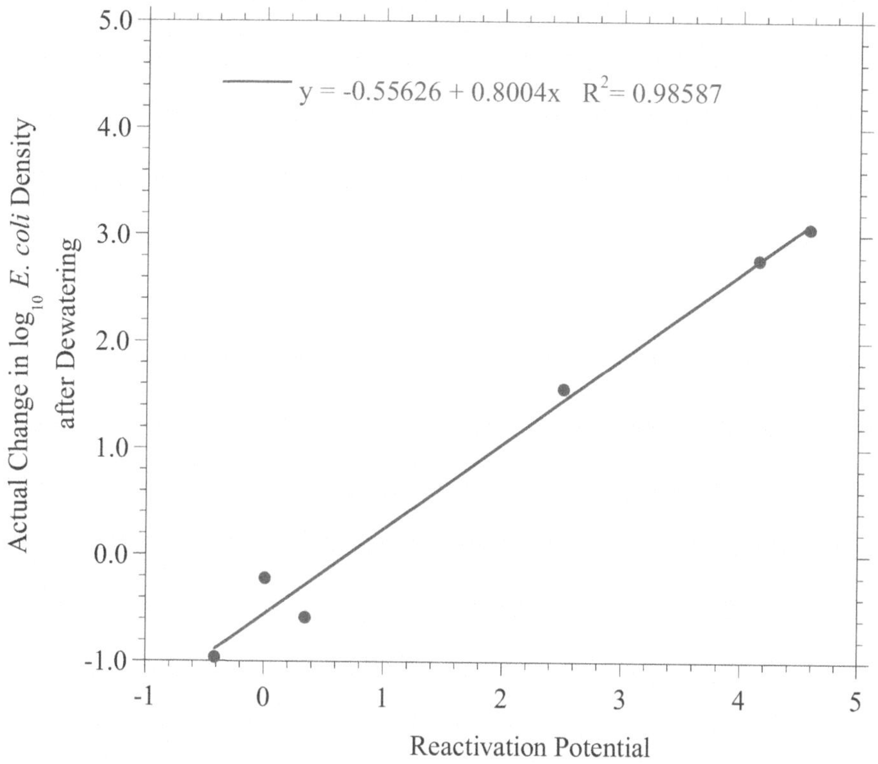

Figure 7-6. Correlation Between Reactivation Potential and Actual Change in the $\log_{10}$ *E. coli* Change After Dewatering.

There is much debate in the literature on what constitutes a viable bacterial cell and when it can be considered dead. As a direct extension of that complexity, measures of viability are also debated, and often functionally defined. For example, measures of viable cells have included, presence of metabolic activity, the ability to replicate, the live/dead stain, culturability, ability to be resuscitated, and quantitative PCR. For the purpose of this research, an appropriate definition of viability may be the ability to be reactivated after dewatering or other means. This would indicate the cells have the means to repair possible injuries, and commence growth and eventually reproduction, even though the bacteria were previously unable to grow on culturing media. As a functional definition, this would also seem most important for regulatory reasons as well as public health issues.

In light of the results from Thermophilic Plant 2 and the time-temperature experiments, it appears that copies of *E. coli* can persist for fairly long time periods. In the case of Thermophilic Plant 2, the *E. coli* DNA enumerated does not appear to represent cells that can be reactivated by dewatering or other perturbations, since significant reactivation did not occur after centrifugation. The potential for reactivation, measured by qPCR densities appeared much greater than actually occurred. This could be due to the nonviable fraction of *E. coli* enumerated by qPCR. For example, Figure 7-6 shows that about 80% of the reactivation potential is realized, suggesting that 20% of the *E. coli* DNA may represent non-viable cells. This could be a significant number in some cases, for example, 20% of a 10^5 cells/g DS would be 20,000 *E. coli* per gram DS.

The literature indicates that cells can become nonculturable when exposed to stress such as low substrate, yet they remain active, with the potential to be resuscitated or to cause infection. However, as exposure to stress continues, the ability to be resuscitated begins to decrease, and activity also ceases, and they could then be considered nonviable. This would indicate that the bacteria cannot endure the stress indefinitely. Therefore, it is possible that the *E. coli* measured in the study represent RNC bacteria for some period of time (or combination of time and temperature), but after that time, the bacteria transition into non-viable stage, incapable of being resuscitated or reactivated. Determining the cutoff for that transition is key for designers of digestion processes.

Based on the evidence collected as part of this research and reported in the literature, the PIs of this report believe that the sudden increase after dewatering can best be explained by the reactivation of non-culturable bacteria theory. Like all scientific theories, by nature, the proof is not definitive, but examination of the outcomes and mechanistic studies, as well as the vast literature on non-culturable organisms, this concept appears to best explain the data. The reasons for non-culturability are not clear, and could include sub-lethal injury, viable but non-culturable cells, presence of inhibitors, or even ingestion by protozoa. Why the bacteria become culturable after dewatering, especially centrifuge dewatering has not been studied, but the literature indicates that first subjecting non-culturable bacteria to specific growth conditions prior to enumeration on standard culturing media allows for resuscitation of the bacteria to a culturable state. It is possible that the cake provides these conditions to support later growth on media. An interesting study by Unc et al. (2006) found that adding sterilized biosolids to soil greatly enhanced the recovery of *E. coli* from soils. Therefore, biosolids may provide specific growth promoters that are able to resuscitate the bacteria. Additional research is needed to further explore these possibilities and investigate more fully the mechanism involved in this phenomenon to either support or refute these theories.

7.3 Evaluation of Mitigation Options for the Sudden Increase in FC

Several utilities have undertaken work to examine methods to mitigate SI and regrowth. Most of this work has examined post digestion modifications, since this would require the least amount of capital cost, in addition, not enough is known about digestion to be assured of successful design modification in the digestion process. The data available on the mitigation options are provided below.

Iranpour et al. (2002 and 2003) reported that SI and regrowth of FC occurred after high solids centrifugation of a thermophilically digested biosolids. Prior to dewatering, the biosolids met Class A requirements of less than 1000 MPN/g DS of FC, but after dewatering using centrifuges, significant increases were measured in the FC density. They surmised that a temperature drop in the sludge in the post-digestion treatment train might be the cause of the SI and regrowth along with possible contamination from the post-digestion treatment process train. The temperature decreased from about 53.5°C leaving digestion to 41°C in the cake storage silo, and was about 40.3°C at the land application farm. Since incubation of FC occurs at 44.5°C during enumeration in the SCM, the decrease in temperature was thought to allow FC to grow in cake. They then insulated and heat traced the post-digestion treatment train, and demonstrated that the temperature only decreased to 50.8°C at the farm site, and no increase in FC was measured, even at the land application site (Iranpour et al., 2006). Iranpour et al. (2006) also stated that this change occurred at the same time as conversion of all the mesophilic digestion to thermophilic, which may have also been responsible for the elimination of the SI.

Erdal et al. (2003) reported that SI and regrowth occurred from a mesophilic digested solids after both low solids and high solids centrifugation, however, after belt filter press dewatering, no SI occurred, and moderate regrowth after one day of storage. Interestingly, they also found that screw conveyance of the cake increased the SI and regrowth of the cake. Erdal et al. (2003 and 2004) found that a low dose lime application to the cake controlled SI and regrowth, especially when the pH was increased to greater than 8.5. They considered low dosages in the range of 1-9% lime on a dry solids basis, and the lime was added prior to the screw conveyance system.

In the case of a TPAD process that met Class A requirements after digestion, but not after centrifuge dewatering, a small continuous feed of 15 ppm (v/v) hypochlorite in digested solids downstream of the high solid centrifuge destroyed a majority of microorganisms and produced a Class B biosolids. This is of considerable interest, since this small dose was not expected to cause much harm to these microorganisms. It appears as though the bacteria were in a very active phase of metabolism and division and therefore more susceptible to low levels of chlorine. This would indicate regrowth/SI rather than cross-contamination.

CHAPTER 8.0

SUMMARY AND RECOMMENDATIONS

8.1 Summary

A summary of the results from this study for SI and regrowth of FC and *Salmonella* is provided in Table 8-1 below. Based on the results from this work, the following summary and conclusions can be made:

- The sudden increase in FC density after dewatering is a real, statistically verified occurrence at some plants. The sudden increase in FC is typically greater in thermophilic digestion processes (up to 4-5 $\log_{10}$ increase) compared to mesophilic digestion processes (0-1 $\log_{10}$ increase). Additional increases in FC densities during cake storage can also occur, with increases from 0-2.5 $\log_{10}$.
- Pathogenic bacteria, such as *Salmonella,* did not "suddenly increase" or regrow in the three Class A biosolids sampled. *Salmonella* did regrow in two Class B biosolids sampled after centrifuge dewatering.
- An evaluation of the data from this study and the literature revealed that about 30 out 35 plants that had the combination of digestion and centrifuge dewatering, observed SI and/or regrowth. Not all the results reported in the literature had analyzed for regrowth, so the proportion could be higher if regrowth was occurring. In contrast, for plants with the combination of digestion and belt filter press dewatering, only about 1 out of 8 had SI and/or regrowth. In this study, a thermophilic plant with reactors in series was the only plant with centrifuge dewatering not to have SI and/or regrowth.
- The results show that dewatering is an important component for SI and regrowth, even more so than digestion.
- Analysis of the results from this work and the literature suggests that the sudden increase in FC is due to the presence of reversibly non-culturable bacteria present after digestion that are reactivated and become culturable after dewatering. Additional research is needed to confirm this hypothesis and also examine other potential causes such as centrifuge contamination.
- The increase during cake storage is due to regrowth of the bacteria in the cake which provides a supportive growing environment. Regrowth can occur even when SI does not occur, such as in the case of mesophilic plants which typically have lower amounts (or no) SI.
- Densities of *E. coli* enumerated using qPCR often were greater than densities measured by standard culturing methods in samples taken after digestion. After centrifuge dewatering, the

increase in the SCM measured densities often resulted in densities very similar to the qPCR measured densities.

- The reactivation potential after digestion, measured as the difference between the qPCR and SCM measured densities, was a good predictor of the actual amount of reactivation measured after dewatering for the plants in this study.
- An evaluation of plant operational parameters showed that digestion temperature was linearly correlated to the reactivation potential, in that as digestion temperature increased, the reactivation potential increased. Digester SRT and VS reduction were not well correlated to the reactivation potential.
- The results suggest that the bacteria can become reversibly non-culturable (RNC) after digestion, and dewatering can reactivate the bacteria, making them culturable again. This would explain the large increase after dewatering. As temperature increases, more of the bacteria become RNC, which can contribute to the greater amounts of reactivation that were observed.
- Similarly, regrowth during cake storage, on average, was greater for thermophilic digestion processes compared to mesophilic digestion processes. For mesophilic digestion, the additional growth measured during cake storage was typically between 0 and 2.5 $\log_{10}$. For thermophilic digestion, the increase was between 2 and 3.5 $\log_{10}$ in the FC densities.
- Laboratory time-temperature experiments generally supported results from the field, in that at higher digestion temperatures, greater numbers of RNC bacteria can develop.

Table 8-1. Summary of SI and Regrowth of FC and *Salmonella* Results.

Plant	Digestion			Dewatering	FC SI	FC Regrowth	*Salmonella* SI	*Salmonella* Regrowth
	Series/ Parallel	Time/ Temp	Class A/B					
Thermo 1	CSTR in Series	15-20 days 55°C	B	HS Centrifuge	Yes	Yes	No	No
Thermo 2	4 CSTR in Series	19-24 days 55°C	A	HS Centrifuge	No	No	No	No
Thermo 3 TPAD	Thermo-Meso in Series	15/20 days 58°C	A	HS Centrifuge	Yes	Yes	No	No
Pre-Past.	Pre-Past. – Meso	19 days 66/35°C	A	HS Centrifuge	Yes	Yes	No	No
Meso 1	Parallel	32 day 38°C	B	HS Centrifuge	No	Yes		
Meso 2	Series	25-30 days 36°C	B	BFP	No	No		
Meso 3	Parallel	22 days 36°C	B	HS Centrifuge	Yes	Yes	No	Yes
Meso 4	Series	70 days 37°C	B	BFP	No	No		
Meso 5	Series	24 days 40°C	B	BFP	No	No		
Meso 6			B	BFP Centrifuge		Yes Yes		Yes No

8.2 Recommendations

The research conducted thus far has provided some important information regarding the issue of reactivation and regrowth. Since the implications of the work are important for the field, several recommendations include:

- Development of mitigation strategies to eliminate issues with reactivation and/or regrowth that can be used by Class B plants. This could include addition of additives before or after dewatering, and ideally provide a cost-effective, simple to implement solution for controlling reactivation and/or regrowth.
- Development of a laboratory based method to enumerate RNC indicator bacteria. This method would help utilities trial approaches to controlling reactivation and/or regrowth at the laboratory scale before going to the full-scale. In addition, the method could help utilities predict how process operational changes may affect reactivation and/or regrowth.
- Perform research to determine if other indicator organisms may be more appropriate and can be used a surrogate for pathogens. For example, this research showed the *Salmonella* did not reactivate/regrow after centrifuge dewatering in the thermophilic processes, yet, FC and *E. coli* did. Therefore, the use of FC as an indicator, especially for Class A plants may not be appropriate, although it may be a conservative indicator. Additional research is needed to better develop appropriate indicators and/or surrogates for pathogens.
- Perform time-temperature studies to define a time-temperature curve for which reactivation potential is eliminated. This will be of importance to the design of future systems, especially ones that do not have reactivation and/or regrowth.

REFERENCES

Adams, B.L., Bates, T.C., and Oliver, J.D. 2003. Survival of *Helicobacter pylori* in a Natural Freshwater Environment. *Appl. Environ. Microbiol.,* **69**, 7462-7466.

Ahn S.J., Costa, J., and Emanuel J.R. 1996. PicoGreen Quantitation of DNA: Effective Evaluation of Samples Pre- or Post-PCR. *Nucleic Acids Res.,* **24**, 2623-2625.

American Public Health Association. 1998. Standard Methods for Examination of Water and Wastewater; 20th Edition. Eds. L.S. Clesceri, A.E. Greenberg, and A.D. Eaton. American Public Health Association, Washington, D.C.

Asakura, H., Watarai, M., Shirahata, T., and Makino, S. 2002. Viable but Nonculturable *Salmonella* Species Recovery and Systemic Infection in Morphine-Treated Mice. *J. Infect. Dis.,* **186**, 1526-1529.

Baker, R.M., Singleton, F.L., and Hood, M.A. 1983. Effects of Nutrient Deprivation on *Vibrio cholerae. Appl. Environ. Microbiol.,***46**, 930-940.

Barry, V.C., Conalty, M.L., Denneny, J.M., and Winder, F. 1956. Peroxide Formation in Bacteriological Media. *Nature,* **178**, 596-597.

Berlin, D.L., Herson, D.S., Hicks, D.T., and Hoover, D.G. 1999. Response of Pathogenic *Vibrio* Species to High Hydrostatic Pressure. *Appl. Environ. Microbiol.,***65**, 2776-2780.

Bhat, H.K. and Epelboym, I., 2004. Quantitative analysis of total mitochondrial DNA: competitive polymerase chain reaction versus real-time polymerase chain reaction. *J Biochem Mol Toxicol.,* **18,** 180-186.

Bogosian, G., Aardema, N.D., Bourneuf, E.V., Morris, P.J.L., and O'Neil, J.P. 2000. Recovery of Hydrogen Peroxide-Sensitive Culturable Cells of *Vibrio vulnificus* Gives the Appearance of Resuscitation from a Viable but Nonculturable State. *Appl. Environ. Microbiol.,* **182**, 5070-5075.

Bogosian, G., Morris, P.J.L., and O'Neil, J.P. 1998. A Mixed Culture Recovery Method Indicates that Enteric Bacteria Do Not Enter the Viable but Nonculturable State. *Appl. Environ. Microbiol.*, **64,** 1736-1742.

Brandl, M.T., Rosenthal, B.M., Haxo, A., and Berk, S.G. 2005 Enhanced survival of *Salmonella enterica* in vesicles released by a soilborne *Tetrahymena* species. *Appl. Environ. Microbiol.* 71, 1562–1569.

Bujoczek, G,R., Reimers, S., and Oleszkiewicz, J.A. 2001b. Abiotic factors affecting inactivation of pathogens in sludge. *Wat. Science & Technol.* 44(10):79-84.

Burge, W.D., Millner, P.D., and Enkiri, N.K. 1986. Regrowth of *Salmonellae* in Composted Sewage Sludge. U.S. EPA Report No. EPA/600/2-86/106.

Byrd, J.J., Xu, H., and Colwell, R.R. 1991. Viable but Nonculturable Bacteria in Drinking Water. *Appl. Environ. Microbiol.,* **57**, 875-878.

Cacho Rivero, J.A. 2005. Anaerobic Digestion of Excess Municipal Sludge: Optimization for Increased Solids Destruction. Ph.D. Thesis, Univ. of Cincinnati.

Cappelier, J.M., Minet, J., Magras, C., Colwell, R.R., and Federighi, M. 1999. Recovery in Embryonated Eggs of Viable but Nonculturable *Campylobacter jejuni* Cells and Maintenance of Ability To Adhere to HeLa Cells after Resuscitation. *Appl. Environ. Microbiol.,* **65**, 5154-5157.

Caro, A., Got, P., Lesne, J., Binard, S., and Baleux, B. 1999. Viability and Virulence of Experimentally Stressed Nonculturable *Salmonella typhimurium. Appl. Environ. Microbiol.,* **65**, 3229-3232.

Chaiyanan, S., Chaiyanan, S., Huq, A., Maugel, T., and Colwell, R.R. 2001. Viability of the Nonculturable *Vibrio cholerae* O1 and O139. *System. Appl. Microbiol.,* **24**, 331-341.

Chaveerach, P., ter Huurne, A.A.H.M., Lipman, L.J.A., and van Knapen, F. 2003. Survival and Resuscitation of Ten Strains of *Campylobacter jejuni* and *Campylobacter coli* under Acid Conditions. *Appl. Environ. Microbiol.,* **69**, 711-714.

Chen, Y.C., Higgins, M.J., Maas, N.A., and Murthy, S.N. 2006. DNA extraction and *Escherichia coli* quantification of anaerobically digested biosolids using the competitive touchdown PCR method. *Water Research*, **40**(16): 3037-3044.

Chen, Y.C., Higgins, M.J., Beightol, S.M., Araujo, G.G., Murthy, S.N., Barben, E.J., and. Toffey, W.E. (2007) Odour Generation and Pathogen Indicator Regrowth after Dewatering: Are They Related? *Proceedings of Int. Wat. Assoc. Specialist Conference-Wastewater Biosolids Sustainability,* Moncton, Canada.

Cheung, K.H.M., Lang, N.L. and Smith, S.R. 2003. The effects of centrifugation dewatering on Escherichia coli numbers in digested sewage sludge. Proceeding of the Joint CIWEM Aqua Enviro Technology Transfer 8th European Biosolids and Organic Residuals Conference, 24 – 26 November, Wakefield.

Colwell, R.R., Brayton, P., Herrington, D., Tall, B., Huq, A., and Levine, M.M. 1996. Viable but Non-Culturable *Vbrio cholerae* O1 Revert to A Cultivable State in the Human Intestine. *World J. Microbiol. Biotechnol.,* **12**, 28-31.

Colwell, R.R., Brayton, P.R., Grimes, D.J., Roszak, D.B., Huq, S.A., and Palmer, L.M. 1985. Viable but Non-Culturable *Vibrio cholerae* And Related Pathogens in the Environment: Implications for Release of Genetically Engineered Microorganisms. *Biotechnol.,* **3**, 817-820.

De Luca, G., Zanetti, F., Fateh-Moghadm, P., and Stampi, S. Sep. 1998. Occurrence of *Listeria monocytogenes* in sewage sludge. *Zentralbl Hyg Unweltmed*. 201(3):269-77.

Demaneche, S., Jocteur-Monrozier, L., Quiquampoix, H., and Simonet, P.. 2001. Evaluation of Biological and Physical Protection against Nuclease Degradation of Clay-Bound Plasmid DNA. *Appl. Environ. Microbiol.*, **67**, 293-299.

Erdal, Z.K., Mendenhall, T.C., Neely, S.K., Wagoner, D.L., and Quigley, C. 2003. Implementing Improvements in a North Carolina Residuals Management Program. *Proceedings of WEF/AWWA/CWEA Joint Residuals and Biosolids Management Conference 2003*, Baltimore, Maryland.

Erdal, Z.K., Wagoner, D.L., Mendenhall, T.C., Neely, S.K., and Quigley, C. 2004. Maintaining Class B Biosolids Post-Dewatering Through Low-Level Lime Dosing. *Proceedings of WEFTEC 2004*, New Orleans, LA.

Fibi, M.R., Broke, M., Schulz, R., Johannsen, R., and Zettlemeissl, G. 1991. Inactivation of recombinant plasmid DNA from a human erythropoietin producing mouse cell line grown on a large scale. *Appl. Microbiol. Biotechnol.*, **35**, 622-630.

Fukushima, H., Tsunomori, Y. and Seki, R. 2003. Duplex real-time SYBR green PCR assays for detection of 17 species of food- or waterborne pathogens in stools. *Journal of Clinical Microbiology*. **41**, 5134-5146.

Geuenich, H.H. and Muller, H.E. 1984. Isolation and germ count of *Listeria monocytogenes* in raw and biologically treated waste water. *Zentralbl Bakteriol Mikrobiol Hyg* [B]. 179:266-73.

Grant, M.A., Weagant, S.D., and Feng, P. 2001. Glutamate Decarboxylase Genes as a Prescreening Marker for Detection of Pathogenic *Escherichia coli* Groups. *Appl. Environ. Microbiol.*, **67**, 3110-3114.

Green, B.L., Clausen, E.M., and Litsky, W. 1977. Two-Temperature Membrane Filter Method for Enumerating Fecal Coliform Bacteria from Chlorinated Effluents. *Appl. Environ. Microbiol.*, **33**, 1259-1264.

Gregersen, P, Grunnet, K., Uldum, S.A., Anderson, B.H., and Madsen, H. 1999. Pontiac fever at a sewage treatment plant in the food industry. *Sc and J Work Environ Health*. 25(3):291-5.

Grey, B. and Steck, T.R. 2001. Concentrations of Copper Thought To Be Toxic to *Escherichia coli* Can Induce the Viable but Nonculturable Condition. *Appl. Environ. Microbiol.*, **67**, 5325-5327.

Gupte, A.R., de Rezende, C.L.E., and Joseph, S.W. 2003. Induction and Resuscitation of Viable but Nonculturable *Salmonella enterica* Serovar Typhimurium DT104. *Appl. Environ. Microbiol.*, **69**, 6669-6675.

Haas, C.N., Rose, J.B. and Gerba, C.P. 1999. Quantitative Microbial Risk Assessment. John Wiley & Sons, Inc. New York.

Hara-Kudo, Y., Ikedo, M., Kodaka, H., Nakagawa, H., Goto, K., Masuda, T., Konuma, H., Kojima, T., and Kumagai, S. 2000. Selective Enrichment with a Resuscitation Step for Isolation of Freeze-Injured *Escherichia coli* O157:H7 from Foods. *Appl. Environ. Microbiol.,* **66**, 2866-2872.

He, J.W. and Jiang, S. 2005. Quantification of Enterococci and Human Adenoviruses in Environmental Samples by Real-Time PCR. *Appl. Environ. Microbiol.*, **71**, 2250-2255.

Heim, S., Lleò, M.D.M., Bonato, B., Guzman, C.A., and Canepari, P. 2002. The Viable but Nonculturable State and Starvation Are Different Stress Responses of *Enterococcus faecalis,* as Determined by Proteome Analysis. *J. Bacterioil.,* **184**, 6739-6745.

Hendrickson, D., Denard, D., Farrell, J., Higgins, M., Murthy, S. *2004.* Reactivation of Fecal Coliforms after Anaerobic Digestion and Dewatering. *Proceedings Water Env. Federation Annual Biosolids and Residuals Conference,* Salt Lake City, Utah.

Hendrickson, D.A. (2001) "HML Biosolids report for a client with thermophilic anaerobic digestion and high solid centrifugation."

Hendrickson, D.A., Denard, D., Farrell, J., Higgins, M., and Murthy, S. 2004. Reactivation of fecal coliforms after anaerobic digestion and dewatering. *Proceedings* 18th Annual Residuals and Biosolids Conference, Salt Lake City, Utah.

Higgins, M.J., Yarosz, D.P., Chen, Y.C., Murthy, S.N., Maas, N., Cooney, J., Glindemann, D., and Novak J.T. 2005. Cycling of volatile organic sulfur compounds in anaerobically digested biosolids and its implications for odors. In Press, *Water Env. Research.*

Higgins, M.J., Murthy, S.N., and Chen, Y.C. (2006) Understanding Factors Affecting Conditioning and Dewatering. Water Environment Research Foundation Report No. 01-CTS-1, Alexandria, VA.

Higgins, M.J., Chen, Y.C., Murthy, S.N., and Hendrickson, D. (2006) Examination of Reactivation of Fecal Coliforms in Anaerobically Digested Biosolids. Water Environment Research Foundation Report No. 03-CTS-13T, Alexandria, VA.

Higuchi, R., Fockler, C., Dollinger, G., and Watson, R. 1993. Kinetic PCR analysis: real-time monitoring of DNA amplification reactions. *Biotechnology (N Y).* **11,** 1026-1030.

Iranpour, R., Cox, H.H.J., Hernandez, G, Redd, K., Fan, S., Moghaddam, O, Abkian, V., Mundine, J., Haug, R.T., and Kearney, R.J. 2003. Production of EQ Biosolids at Hyperion Treatment Plant: Problems and Solutions for Reactivation/Growth of Fecal Coliforms. *Proceedings of WEFTEC 2003*, Los Angeles.

Iranpour, R., Oh, S., Cox, H.J., Samar, P., Taylor, D., Mohamed, F., Hagekhalil, A, and Kearney, R.J. 2002. Effects of Dewatering on Bacterial Inactivation: Centrifuge Simulation and Field Tests at the Hyperion Treatment Plant. *Proceedings of WEFTEC 2002.*

Jolis, D. (2006) Regrowth of Fecal Coliforms in Class A Biosolids. *Water Env. Research*, **78**, 442-446.

King, C.H., Shotts Jr., E.B., Wooley, R.E. and Porter, K.G. 1988 Survival of coliforms and bacterial pathogens within protozoa during chlorination. *Appl. Environ. Microbiol.* 54, 3023–3033.

Kolling, G. and Matthews, K.R. 2001. Examination of Recovery In Vitro and In Vivo of Nonculturable *Escherichia coli* O157:H7. *Appl. Environ. Microbiol.,* **67**, 3928-3933.

Lebuhn, M., Effenberger, M., Garces, G., Gronauer, A., and Wilderer, P.A.. (2005) Hygenization by anaerobic digestion: comparison between evaluation by cultivation and quantitative real-time PCR. *Wat. Sci. Technol.*, **52**, 93-99.

LeChevallier, M.W., Cameron, S.C., and McFeters, G.A. 1983. New Medium for Improved Recovery of Coliform Bacteria from Drinking Water. *Appl. Environ. Microbiol.,* **45**, 484-492.

LeChevallier, M.W., Singh, A., Schiemann, D.A., and McFeters, G.A. 1985. Changes in Virulence of Waterborne Enteropathogens with Chlorine injury. *Appl. Environ. Microbiol.,* **50**, 412-419.

Lin, S.D. 1976. Membrane Filter Method for Recovery of Fecal Coliforms in Chlorinated Sewage Effluents. *Appl. Environ. Microbiol.,* **32**, 547-552.

Lleò, M.D.M., Bonato, B., Tafi, M.C., Signoretto, C., Boaretti, M., and Canepari, P. 2001. Resuscitation rate of different enterococcal species in the viable but non-culturable state. *J. Appl. Microbiol.*, **91**, 1095-1102.

Lleò, M.D.M., Tafi, M.C., and Canepari, P. 1998. Nonculturable *Enterococcus faecalis* Cells Are Metabolically Active and Capable of Resuming Active Growth. *System. Appl. Microbiol.,* **21**, 333-339.

Lorenz, M.G. and Wackernagel, W. 1987. Adsorption of DNA to Sand and Variable Degradation Rates of Adsorbed DNA. *Appl. Environ. Microbiol.*, **53**, 2948-2952.

Lorenz, M.G. and Wackernagel, W. 1994. Bacterial Gene Transfer by Natural Genetic Transformation in the Environment. *Microbiol. Rev.*, **58**, 563-602.

Lucero-Ramirez, B. and Malina, J.F. 2000. The Effects of Time and Temperature on the Fate of Pathogens and Indicator Bacteria During Municipal Wastewater Sludge – Mesophilic Anaerobic Digestion, Air Drying, and Composting. Ph.D. Dissertation, University of Texas, Austin..

Makino, S., Kii, T., Asakura, H., Shirahata, T., Ikeda, T., Takeshi, K., and Itoh, K. 2000. Does Enterohemorrhagic *Escherichia coli* O157:H7 Enter the Viable but Nonculturable State in Salted Salmon Roe? *Appl. Environ. Microbiol.,* **66**, 5536-5539.

Martin, S.E., Flowers, R.S., and Ordal, Z.J. 1976. Catalase: Its Effect on Microbial Enumeration. *Appl. Environ. Microbiol.,* **32**, 731-734.

McDaniels A.E., Rice, E.W., Reyes, A.L., Johnson, C.H., Haugland, R.A., and Stelma, G.N., Jr. 1996. Confirmational identification of *Escherichia coli*, a comparison of genotypic and phenotypic assays for glutamate decarboxylase and beta-D- glucuronidase. *Appl. Environ. Microbiol.*, **62**, 3350-3354.

McDonald, L.C., Hackney, C.R., and Ray, B. 1983. Enhanced Recovery of Injured *Escherichia coli* by Compounds That Degrade Hydrogen Peroxide or Block Its Formation. *Appl. Environ. Microbiol.,* **45**, 360-365.

McFeters, G.A., Cameron, S.C., and LeChevallier, M.W. 1982. Influence of Diluents, Media, and Membrane Filters on Detection of Injured Waterborne Coliform Bacteria. *Appl. Environ. Microbiol.,* **43**, 97-103.

McFeters, G.A., Kippin, J.S., and LeChevallier, M.W. 1986. Injured Coliforms in Drinking Water. *Appl. Environ. Microbiol.,* **51**, 1-5.

Meckes, M. 2006. Personal communication. Mark Meckes, EPA, Cincinnati, Ohio.

Meckes, M. and E. Rhodes. 2004. Evaluation of bacteriological indicators and disinfection for alkaline treated biosolids. *J. Env. Eng. Sci.*, **3**, 231-236.

Mizunoe, Y., Wai, S.N., Takade, A., and Yoshida, S. 1999. Restoration of Culturability of Starvation-Stressed and Low-Temperature-Stressed *Escherichia coli* O157 Cells by Using H_2O_2-Degrading Compounds. *Arch Microbiol.,* **172**, 63-67.

MMWR. Legionnaires' Disease Associated with Potting Soil—California, Oregon, and Washington, May-June 2000. United States, 2000. 49(34):777-778.

Moellering, 1992). Emergence of *Enterococcus* as a significant pathogen. Clin. Infect. Dis. **14:**1173–1176.

Molloy, M.P., Herbert, B.R., Slade, M.B., Rabilloud, T., Nouwens, A.S., Williams, K.L., and Gooley, A.A. 2000. Proteomic Analysis of the *Escherichia coli* Outer Membrane. *Eur. J. Biochem.,* **267**, 2871-2881.

Monteleone, M.C., Furness, D., Jefferson, B., and Cartmell, E. 2004. Fate of *E. coli* Across Mechanical Dewatering Processes. *Env. Technol.,* **25**, 825-831.

Murthy, S.N., Higgins, M.J., and Chen, Y.C. 2003. The Effect of Dewatering Equipment on VSC and Odor Production from Digested Biosolids. *Proceedings of Water Env. Fed. 76th Annual Conf.*, Los Angeles.

NAS 2001. "Biosolids Applied to Land: Advancing Standards and Practices" Committee on Toxicants and Pathogens in Biosolids Applied to Land, National Research Council.

Oliver, J.D. 2005. The Viable but Nonculturable State in Bacteria. *J. Microbiology*, **43**, 93-100.

Oliver, J.D. 1995. The Viable but Non-Culturable State in the Human Pathogen *Vibrio vulnificus*. *FEMS Microbiol. Lett.,* **133**, 203-208.

Pfeffer, J.M., Strating, H., Weadge, J.T., and Clarke, A.J. 2006. Peptidoglycan O Acetylation and Autolysin Profile of *Enterococcus faecalis* in the Viable but Nonculturable State. *J. Bacteriology*, **188**, 902-908.

Phillips, S.J., Dalgarn, D.S., and Young, S.K. 1989. Recombinant DNA in wastewater: pBR322 degradation kinetics. *J. Wat. Poll. Control Fed.*, **61**, 1588-1595.

Pommepuy, M., Butin, M., Derrien, A., Gourmelon, M., Colwell, R.R., and Cormier, M. 1996. Retention of Enteropathogenicity by Viable but Nonculturable *Escherichia coli* Exposed to Seawater and Sunlight. *Appl. Environ. Microbiol.,* **62**, 4621-4626.

Qi, Y.N., Gillow, S., Herson, D.S., Dentel, S.K. 2004. Reactivation and/ or growth of fecal coliform bacteria during centrifugal dewatering of anaerobically digested biosolids. *Wat. Sci. Technol.,* **50**, 115-120.

Raeymaekers, L. 1993. Quantitative PCR: theoretical considerations with practical implications. *Anal. Biochem.,* **214**, 582-585.

Raeymaekers, L. 1995. A commentary on the practical applications of competitive PCR. *Genome Res.,* **5**, 91-94.

Rahman, I., Shahamat, M., and Chowdhury, M.A.R. 1996. Potential Virulence of Viable but Nonculturable *Shigella dysenteriae* Type 1. *Appl. Environ. Microbiol.,* **62**, 115-120.

Rahman, M.H., Suzuki, S., and Kawai, K. 2001. Formation of viable but non-culturable state (VBNC) of *Aeromonas hydrophila* and its virulence in goldfish, *Carassius auratus*. Microbiological Research 156(1):103-106.

Reissbrodt, R., Heier, H., Tschäpe, H., Kingsley, R.A., and Williams, P.H. 2000. Resuscitation by Ferrioxamine E of Stressed *Salmonella enterica* Serovar Typhimurium from Soil and Water Microcosms. *Appl. Environ. Microbiol.,* **66**, 4128-4130.

Reissbrodt, R., Rienaecker, I., Romanova, J.M., Freestone, P.P.E., Haigh, R.D., Lyte, M., Tschäpe, H., and Williams, P.H. 2002. Resuscitation of *Salmonella enterica* Serovar Typhimurium and Enterohemorrhagic *Escherichia coli* from the Viable but Nonculturable State by Heat Stable Enterobacterial Autoinducer. *Appl. Environ. Microbiol.,* **68**, 4788-4794.

Rockabrand, D., Austin, T., Kaiser, R., and Blum, P. 1999. Bacterial Growth State Distinguished by Single-Cell Protein Profiling: Does Chlorination Kill Coliforms in Municipal Effluent? *Appl. Environ. Microbiol.,* **65**, 4181-4188.

Romprè A., Servais, P., Baudart, J., de-Roubin, M., and Laurent, P. 2002. Detection and enumeration of coliforms in drinking water: current methods and emerging approaches. *J. Microbiol. Methods,* **49**, 31-54.

Ronchi, E. and Wald, S. 1999. New Molecular Technologies for Safe Drinking Water. *Observer,* **215**,

Rose, P., Harkin, J.M, and Hickey, W.J. 2003. Competitive touchdown PCR for estimation of *Escherichia coli* DNA recovery in soil DNA extraction. *J. Microbiol. Methods,* **52**, 29-38.

Ruiz, T.R., Andrews, S., and Smith, G.G.. 2000. Identification and characterization of nuclease activities in anaerobic environmental samples. *Canadian J. Microbiol.*, **46**, 736-740.

Signoretto, C., Lleò, M.D.M., Tafi, M.C., and Canepari, P., 2000. Cell Wall Chemical Composition of *Enterococcus faecalis* in the Viable but Nonculturable State. *Appl. Environ. Microbiol.,***66**, 1953-1959.

Stuart, D.G., McFeters, G.A., and Schillinger, J.E. 1977. Membrane Filter Technique for the Quantification of Stressed Fecal Coliforms in the Aquatic Environment. *Appl. Environ. Microbiol.,* **34**, 42-46.

Sun, N.W., Mizunoe, Y., and Takada, A. 2000. A comparison of solid and liquid media for resuscitation of starvation- and low-temperature-induced nonculturable cells of *Aeromonas hydrophila*. Arch. Microbiol. 173:307-310.

Tezcan-Merdol, D., Ljungstrom, M., Winiecka-Krusnell, J., Linder, E., Engstrand, L. & Rhen, M. 2004 Uptake and replication of *Salmonella enterica* in *Acanthamoeba rhysodes*. *Appl. Environ. Microbiol.* 70, 3706–3714.

Tholozan, J.L., Cappelier, J.M., Tissier, J.P., Delattre, G., and Federighi, M. 1999. Physiological Characterization of Viable-but-Nonculturable *Campylobacter jejuni* Cells. *Appl. Environ. Microbiol.,* **65**, 1110-1116.

Unc, A., J. Gardner, and S. Springthorpe. 2006. Recovery of *Escherichia coli* from Soil after Addition of Sterile Organic Wastes. *Appl. Environ. Microbiol.*, **72**, 2287-2289.

U.S. EPA (1992) Control of Pathogens and Vector Attraction in Sewage Sludges - Under 40 CFR Part 503. EPA/625/R-92/013, Revised October 1999. Washingtion, D.C.

U.S. EPA 1999. "Control of Pathogens and Vector Attraction in Sewage Sludge." EPA/625/R-92/013.

U.S. EPA 2002. "Workshop on Emerging Infectious Disease Agents and Issues Associated with Animal Manures, Biosolids and Other Similar By-Products."

U.S. EPA 2005a. Method 1680: Fecal Coliforms in Sewage Sludge (Biosolids) by Multiple-Tube Fermentation using Lauryl Tryptose Broth (LTB) and EC Medium. EPA-821-R-04-026, Washington, D.C.

U.S. EPA 2005b. Method 1603: *Escherchia coli* (*E. coli*) in Water by Membrane Filtration using Membrane-Thermotolerant *Escherchia coli* Agar (Modified mTEC). EPA-821-R-04-025, Washington, D.C.

U.S. EPA. 2006. *Aeromonas*: Human Health Criteria Document. U.S. EPA Office of Water, published March 2006.

Wai, S.N., Mizunoe, Y., Takade, A., and Yoshida, S.. 2000. A comparison of solid and liquid media for resuscitation of starvation- and low-temperature-induced nonculturable cells of *Aeromonas hydrophila.* Archives of Microbiology 173(4):307-310.

Wang, A.M., Doyle, M.V., and Mark, D.F. 1989. Quantitation of mRNA by the polymerase chain reaction. *Proc. Natl. Acad. Sci. U.S.A.,* **86**, 9717-9721.

Wang, G. and Doyle, M.P. 1998. Survival of Enterohemorrhagic *Escherichia coli* O157:H7 in Water. *J. Food Protection*, **61**, 662-667.

Whitesides, M.D. and Oliver, J.D. 1997. Resuscitation of *Vibrio vulnificus* from the Viable but Nonculturable State. *Appl. Environ. Microbiol.,* **63**, 1002-1005.

Zachar, V., Thomas, R.A., and Goustin, A.S. 1993. Absolute quantification of target DNA: a simple competitive PCR for efficient analysis of multiple samples. *Nucleic Acids Res.,* **21**: 2017-2018.

WASTEWATER UTILITY

Alabama
Montgomery Water Works & Sanitary Sewer Board

Alaska
Anchorage Water & Wastewater Utility

Arizona
Glendale, City of, Utilities Department
Mesa, City of
Peoria, City of
Phoenix Water Services Dept.
Pima County Wastewater Management
Safford, City of

Arkansas
Little Rock Wastewater Utility

California
Central Contra Costa Sanitary District
Corona, City of
Crestline Sanitation District
Delta Diablo Sanitation District
Dublin San Ramon Services District
East Bay Dischargers Authority
East Bay Municipal Utility District
Eastern Municipal Water District
El Dorado Irrigation District
Fairfield-Suisun Sewer District
Fresno Department of Public Utilities
Inland Empire Utilities Agency
Irvine Ranch Water District
Las Virgenes Municipal Water District
Livermore, City of
Los Angeles, City of
Los Angeles County, Sanitation Districts of
Napa Sanitation District
Orange County Sanitation District
Palo Alto, City of
Riverside, City of
Sacramento Regional County Sanitation District
San Diego Metropolitan Wastewater Department, City of
San Francisco, City & County of
San Jose, City of
Santa Barbara, City of
Santa Cruz, City of
Santa Rosa, City of
South Bayside System Authority
South Coast Water District
South Orange County Wastewater Authority
Stege Sanitary District
Sunnyvale, City of
Union Sanitary District
West Valley Sanitation District

Colorado
Aurora, City of
Boulder, City of
Greeley, City of
Littleton/Englewood Water Pollution Control Plant
Metro Wastewater Reclamation District, Denver

Connecticut
Greater New Haven WPCA
Stamford, City of

District of Columbia
District of Columbia Water & Sewer Authority

Florida
Broward, County of
Fort Lauderdale, City of
Miami-Dade Water & Sewer Authority
Orange County Utilities Department
Reedy Creek Improvement District
Seminole County Environmental Services
St. Petersburg, City of
Tallahassee, City of
Tampa, City of
Toho Water Authority
West Palm Beach, City of

Georgia
Atlanta Department of Watershed Management
Augusta, City of
Clayton County Water Authority
Cobb County Water System
Columbus Water Works
Fulton County
Gwinnett County Department of Public Utilities
Savannah, City of

Hawaii
Honolulu, City & County of

Idaho
Boise, City of

Illinois
American Bottoms Wastewater Treatment Plant
Greater Peoria Sanitary District
Kankakee River Metropolitan Agency
Metropolitan Water Reclamation District of Greater Chicago
Wheaton Sanitary District

Iowa
Ames, City of
Cedar Rapids Wastewater Facility
Des Moines, City of
Iowa City

Kansas
Johnson County Unified Wastewater Districts
Unified Government of Wyandotte County/ Kansas City, City of

Kentucky
Louisville & Jefferson County Metropolitan Sewer District
Sanitation District No. 1

Louisiana
Sewerage & Water Board of New Orleans

Maine
Bangor, City of
Portland Water District

Maryland
Anne Arundel County Bureau of Utility Operations
Howard County Department of Public Works
Washington Suburban Sanitary Commission

Massachusetts
Boston Water & Sewer Commission
Massachusetts Water Resources Authority (MWRA)
Upper Blackstone Water Pollution Abatement District

Michigan
Ann Arbor, City of
Detroit, City of
Holland Board of Public Works
Saginaw, City of
Wayne County Department of Environment
Wyoming, City of

Minnesota
Rochester, City of
Western Lake Superior Sanitary District

Missouri
Independence, City of
Kansas City Missouri Water Services Department
Little Blue Valley Sewer District
Metropolitan St. Louis Sewer District

Nebraska
Lincoln Wastewater System

Nevada
Henderson, City of
Las Vegas, City of
Reno, City of

New Jersey
Bergen County Utilities Authority
Ocean County Utilities Authority
Passaic Valley Sewerage Commissioners

New York
New York City Department of Environmental Protection

North Carolina
Charlotte/Mecklenburg Utilities
Durham, City of
Metropolitan Sewerage District of Buncombe County
Orange Water & Sewer Authority

Ohio
Akron, City of
Butler County Department of Environmental Services
Columbus, City of
Metropolitan Sewer District of Greater Cincinnati
Northeast Ohio Regional Sewer District
Summit, County of

Oklahoma
Oklahoma City Water & Wastewater Utility Department
Tulsa, City of

Oregon
Albany, City of
Clean Water Services
Eugene, City of
Gresham, City of
Portland, City of Bureau of Environmental Services
Water Environment Services

Pennsylvania
Hemlock Municipal Sewer Cooperative (HMSC)
Philadelphia, City of
University Area Joint Authority

South Carolina
Charleston Water System
Mount Pleasant Waterworks & Sewer Commission
Spartanburg Water

Tennessee
Cleveland, City of
Knoxville Utilities Board
Murfreesboro Water & Sewer Department
Nashville Metro Water Services

Texas
Austin, City of
Dallas Water Utilities
Denton, City of
El Paso Water Utilities
Fort Worth, City of
Houston, City of
San Antonio Water System
Trinity River Authority

Utah
Salt Lake City Corporation

Virginia
Alexandria Sanitation Authority

Arlington, County of
Fairfax County
Hampton Roads Sanitation District
Hanover, County of
Henrico, County of
Hopewell Regional Wastewater Treatment Facility
Loudoun County Sanitation Authority
Lynchburg Regional WWTP
Prince William County Service Authority
Richmond, City of
Rivanna Water & Sewer Authority

Washington

Everett, City of
King County Department of Natural Resources
Seattle Public Utilities
Sunnyside, Port of
Yakima, City of

Wisconsin

Green Bay Metro Sewerage District
Kenosha Water Utility
Madison Metropolitan Sewerage District
Milwaukee Metropolitan Sewerage District
Racine, City of
Sheboygan Regional Wastewater Treatment
Wausau Water Works

Australia

ACTEW (Ecowise)
South Australian Water Corporation
Sydney Water Corporation
Water Corporation of Western Australia

Canada

Lethbridge, City of
Regina, City of, Saskatchewan
Toronto, City of, Ontario
Winnipeg, City of, Manitoba

New Zealand

Watercare Services Limited

United Kingdom

United Utilities North West (UUNW)

STORMWATER UTILITY

California

Fresno Metropolitan Flood Control District
Los Angeles, City of, Department of Public Works
Monterey, City of
Sacramento, County of
San Francisco, City & County of
Santa Rosa, City of
Sunnyvale, City of

Colorado

Aurora, City of
Boulder, City of

Georgia

Griffin, City of

Iowa

Cedar Rapids Wastewater Facility
Des Moines, City of

Kansas

Overland Park, City of

Kentucky

Louisville & Jefferson County Metropolitan Sewer District

Maine

Portland Water District

North Carolina

Charlotte, City of, Stormwater Services

Pennsylvania

Philadelphia, City of

Tennessee

Chattanooga Stormwater Management

Texas

Harris County Flood Control District, Texas

Washington

Bellevue Utilities Department
Seattle Public Utilities

STATE

Arkansas Department of Environmental Quality
Connecticut Department of Environmental Protection
Kansas Department of Health & Environment
Kentucky Department of Environmental Protection
New England Interstate Water Pollution Control Commission (NEIWPCC)
Ohio River Valley Sanitation Commission
Urban Drainage & Flood Control District, CO

CORPORATE

ADS Environmental Services
Advanced Data Mining International
Alan Plummer & Associates
Alpine Technology Inc.
Aqua-Aerobic Systems Inc.
Aquateam–Norwegian Water Technology Centre A/S
ARCADIS
Associated Engineering
Black & Veatch
Blue Water Technologies, Inc.
Boyle Engineering Corporation
Brown & Caldwell
Burgess & Niple, Ltd.
Burns & McDonnell
CABE Associates Inc.
The Cadmus Group
Camp Dresser & McKee Inc.
Carollo Engineers Inc.
Carpenter Environmental Associates Inc.
CDS Technologies Inc.
CET Engineering Services
Chemtrac Systems Inc.
CH2M HILL
CRA Infrastructure & Engineering
CONTECH Stormwater Solutions
D&B/Guarino Engineers, LLC
Damon S. Williams Associates, LLC
Earth Tech Inc.
Ecovation
EMA Inc.
Environmental Operating Solutions, Inc.
Environ/The ADVENT Group, Inc.
Fay, Spofford, & Thorndike Inc.
Freese & Nichols Inc.
ftn Associates Inc.
Gannett Fleming Inc.
Garden & Associates, Ltd.
Geosyntec Consultants
GHD
Golder Associates Ltd.
Greeley and Hansen LLC
Hazen & Sawyer, P.C.
HDR Engineering Inc.
HNTB Corporation
Hydromantis Inc.
HydroQual Inc.
Infilco Degremont Inc.
Jacobson Satchell Consultants, Inc.
Jacques Whitford NAWE, Inc.
Jason Consultants LLC Inc.
Jordan, Jones, & Goulding Inc.
KCI Technologies Inc.
Kelly & Weaver, P.C.
Kennedy/Jenks Consultants
KMK Consultants
Komline Sanderson Engineering Corporation
Larry Walker Associates
Limno-Tech Inc.
The Low Impact Development Center Inc.
Malcolm Pirnie Inc.
Material Matters
McKim & Creed
Metcalf & Eddy Inc.
Monteco Corporation
MPR Engineering Corporation, Inc.
MWH
NewFields Water Resources, LLC
O'Brien & Gere Engineers Inc.
Odor & Corrosion Technology Consultants Inc.
Original Engineering Consultants, Ltd.
Oscar Larson & Associates
Parametrix Inc.
Parsons
Post, Buckley, Schuh & Jernigan
Praxair, Inc.
Ring Industrial Group
RMC Water & Environment
Ross & Associates Ltd.
Rothberg, Tamburini & Windsor, Inc.
SAIC
Savin Engineers
Siemens Water Technologies
Stantec Consulting Inc.
Stearns & Wheler, LLC
Stone Environmental Inc.
Stratus Consulting Inc.
Synagro Technologies Inc.
Tetra Tech Inc.
Trojan Technologies Inc.
Trussell Technologies, Inc.
URS Corporation
Wade-Trim Inc.
Westin Engineering Inc.
Weston Solutions Inc.
Woodard & Curran
Zenon Environmental Inc./GE Water Process Technology
Zoeller Pump Company

INDUSTRY

American Electric Power
American Water
ChevronTexaco Energy Research & Technology Company
The Coca-Cola Company
Dow Chemical Company
DuPont Company
Eastman Chemical Company
Eli Lilly & Company
Merck & Company Inc.
Premier Chemicals LLC
Procter & Gamble Company
Thames Water Utilities
Severn Trent Services Inc.
Suez Environnment
United Water Services LLC

Note: List as of 10/31/07

IWA Publishing's authorised EU representative for General Product Safety Regulations is Diane D'Arras, 15 rue Duret, 75116 Paris, France, e-mail: safety@iwap.co.uk.

Printed and bound by CPI Group (UK) Ltd, Croydon, CR0 4YY
12/05/2026
02108865-0016